864-752-1673

URANIUM
SEEKERS

URANIUM SEEKERS

*A Photo-Essay Tribute
To Miners*

Craig Evan Royce

AuthorHouse™
1663 Liberty Drive
Bloomington, IN 47403
www.authorhouse.com
Phone: 1-800-839-8640

© 2012 by Craig Evan Royce. All rights reserved.

No part of this book may be reproduced, stored in a retrieval system, or transmitted by any means without the written permission of the author.

Published by AuthorHouse   06/29/2012

ISBN: 978-1-4772-0400-9 (sc)
ISBN: 978-1-4772-0399-6 (hc)
ISBN: 978-1-4772-0325-5 (e)

Library of Congress Control Number: 2012908030

Any people depicted in stock imagery provided by Thinkstock are models, and such images are being used for illustrative purposes only.
Certain stock imagery © Thinkstock.

This book is printed on acid-free paper.

Because of the dynamic nature of the Internet, any web addresses or links contained in this book may have changed since publication and may no longer be valid. The views expressed in this work are solely those of the author and do not necessarily reflect the views of the publisher, and the publisher hereby disclaims any responsibility for them.

To all miners
from Crandall Canyon, Utah
to Upper Big Branch, West Virginia,
and on to China, miners who now
lay, eternally entombed, where
standing, within their portal.

For miners are persons who make light where once there was only
darkness...

and

the Earl Keith Royce Family

"Little Miner Boy"
"Courtesy, Western Mining and
Railroad Museum all rights reserved."
Helper, Utah

Contents

Little Miner Boy vi
About the Photographers ix
Preface xi
Central City, Colorado 1
Temple Mountain, Utah 15
One Such Visionary 27
The Last Miners Of Temple Mountain, Utah 39
Poulet and the Curies 58
Young Forest Ranger 67
1940's Glimpse 77
Zane Grey Lingers 90
First draft outline material and Oral History Materials ... 109
Acknowledgements 143

Martin's Sunset Strip Studio Lobby

About The Photographers

Introducing the chief, contracted, photographer of Uranium Seekers, "artonfilm.org" sums him up thusly, in their description of the DVD, MARTIN: THE HOLLYWOOD YEARS,

"World – famous photographer, Martin, lived in Hollywood between 1956 and 1996. During this period, he captured wonderful images of international stars, aspiring actors, TV personalities, musicians and many other people with interesting faces. His lighting technique for black-and white photos is unparalleled and his composition is highly-acclaimed…"

I first became aware of Martin's work when we were both published in the mid-1970's by the Ward Ritchie Press of Pasadena, California. A fine house who has also published work by Ansel Adams, Dorthea Lange, Robertson Jeffers, Howard Fast, Will Soule, Adam Clark Vroman, and others. Ward Ritchie Press had called my Country Miles are Longer Than City Miles with photos by Jeff Gitlin, "An important document in the art and social history of Americana."

In 1975, when approached in my gallery/museum in Laguna Beach, California by Brenda Migliaccio (Mă - lăt - chi) Kalatzes of Migliaccio Uranium Properties to develop material for a book on her father, Lawrence Migliaccio, and other space-age pioneers of the uranium mining regions of the American west, I immediately thought of Martin and how cool it would be to take him from the "Sunset Strip" to Temple Mountain, Utah adding his work to Ansel Adams, Dorthea Lange, Will Soule, and the others documenting the topography and personalities

Craig Evan Royce

of America's west including portraits of native American Indians from the, then, Uinta and Ouray Tribal Lands of Utah in 1976.

The first wild photographic expedition to Utah's San Rafael Swell, as Brenda had already known Martin, was memorable. Fortunately, the contracted images of this expedition have been maintained by Ms. Migliaccio and exist for this presentation.

Martin and his talented son Peter were not to accompany us on other expeditions, the national and international events surrounding uraniums utilization tempered the publishing industries interest.

We remained friends and even conceptualized another photo-art project with images by Martin, we came close at POCKET BOOKS. Trips to his Sunset Strip studio were always fabulous.

I remember a gourmet meal in Martin's Hollywood Flat once occupied by one of Bugsy Siegel's chief lieutenants, Mickey Cohen, girlfriends. I cannot remember whether it was Tempest Storm, Candy Barr, or Liz Renay, but Martin pointed out the bullet holes in the wall and ceiling the lady fired at Mickey. As a world-famous Hungarian immigrant his stories of the revolution of the mid-1950's were riveting.

Martin told me, when discussing his creative process, Martin knew he had a good photographic image when he could discern his image in the eyes of his subject. Look closely.

Two, moving, photos are credited to the legendary award winning, sports and social photographer Al Szabol.

Preface 2012

When guiding through Zane Grey's, John Ford's, and Edward Abbey's Colorado Plateau regions, and environs, of the western U.S.A., I cannot escape Rene Daumal's description of vast vistas in his short Mountain Climbing novel Mount Analogue, the French metaphysical novelist states,

> "'. . . Either way, Karl is right about our reactions to mountains. Victor Hugo, coming down from the Rigi, which even in his day was not considered very high, remarked that the view of the world from high peaks does such a violence to our visual habits that the natural takes on the appearance of the supernatural. He even asserted that the average human mind cannot bear such a wrenching of its perceptions . . .'"

Thus, it must have been with me when, in 1975, I first ascended Utah's mammoth petrified Jurassic San Rafael Reef and Temple Mountain.

The draft snatches for this document come from a much larger work given various working titles. One draft was over 500 pages which Oxford University Press called, as I recall, awesome but too specialized.

Much of the early material was gleaned from the archives of the great uranium mining pioneer Howard Wilson Balsley of Moab, Utah who graciously afforded me his archives to creatively use as I saw fit. These materials included the half-century old Uranium Country by Kathleen Bruyn. The Migliaccio Uranium Properties archives, State of Utah archives, personal communications, and the decades of mineral resource literature I've accumulated by virtue of mineral rights ownership in the Temple Mountain Mining District, Emery County, Utah for the past three decades, were heavily relied on.

Through the mid-to-late 1970's, while developing material for this photo-essay, I did a tad of administrative work for Migliaccio Uranium Properties and it was here I also began my three decade plus dealings with the U.S. Department of the Interior Bureau of Land Management post Federal Land Policy Management Act of 1976. This primarily dealt with the operating Vanadium King Mines on North Temple Mountain.

In late 1979 and early 1980 I put together the capital to relocate a 12,000 acre land package originally located by Chester Farrow, consulting geologist, his wife Betty, Hub Newell and his wife Maxine in the Temple Mountain Mining District. Previously, they had leased the property to Exxon Corporation.

From that date to the present, 2012, I have maintained some measure of mineral position in the Temple Mountain Mining District while, today, my daughter, Lexington, Kentucky attorney Mackenzie Victoria Royce owns the developmental mineral rights to the Crows Nest Lode and the Black Beauty No. 2 uranium and vanadium mine in the district. These were originally located in 1936.

Some small measure of a few of my years in the Temple Mountain Mining District has been recorded by Joseph M. Bauman's Jr.'s work as environment editor of the Deseret News in Salt Lake City, Utah and his book Stone House Lands.. The San Rafael Reef published by the University of Utah Press.

Strangely, a SENCO-PHENIX INTENSIVE CULTURAL RESOURCE SURVEY . . . Antiquities Project Number U-84-14-313s by John A. Senulis names this book and several archaic and prehistoric sites I recorded to Smithsonian Site Status through the Utah Division of State History in their ARCHAEOLOGICAL POTENTIAL section.

Uranium Seekers

Martin Image "Anna, oldest resident of the
Uinta and Ouray Tribal Lands, Utah, 1976

The essence of the project was to pay tribute to the miners who traversed Zane Grey's and John Ford's great western expanse in search of uranium ore, one rock at a time, from prior to Madame Curies alleged trips. Also to remind the reader that uranium was first used to kill, not humanity, rather neoplasms (cancer),

I harbored the hope that by going back to the first uranium rocks, nuclear industry would re-evaluate the physical structure of nuclear reactors and parks, one cubic yard at a time.

For nuclear reactors have for decades been constructed and often located with too much haste, witness Fukushima Daiichi. Like uranium miners themselves, it is the hands of the humanity who add mortar to the cement in which reactors rest which determines safety.

With a Daumal induced wrenching of my perceptions I hoped uraniums destructive reality would melt under harmonious utilization.

Even with the respect and support of the fine Beverly Hills, California literary agent, Clyde M. Vandeburg of Vandeburg-Linkletter, Associates who, at the time, represented Ronald and Nancy Reagan, Barry Goldwater and so many others, in time Uranium Seekers lost its legs, Martin's unique images were put to bed for decades.

However, I recently learned the "Utah State Historical Quarterly" Unpublished Manuscripts Division of the Department of Community and Culture, at the Utah State Archives, harbored some Uranium Seekers material for decades and the recent events at Fukushima Daiichi made uranium part of the international conversation once again, I decided to dust off Martin's images as well as draft snatches of original text.

As cold winds from the north, and Canada, inch into my weathered, 1942, Defense Plant Corporation, Geneva Mine, miners cottage set against Utah's Book Cliff, Roan Cliff, and West Tavaputs Plateau escarpments protecting Nine-Mile Canyon and conservation easement shielded Range Creek, in Bo Huff inspired East Carbon City, Utah, my minds eye perceives the image of Sego Lily, Professor Don Burge, or my Daughter: one, well dressed, individual figure framed by ripple marked shale and pre-historic pinyon pine, when, the wild rocky mountain big horns are resting on rice grass below the petroglyphs during golden hour.

CENTRAL CITY, COLORADO

The earliest record of uranium's existence dates to the year 1517 at St. Joachimsthal, in Bohemia now The Czech Republic. An oxide of uranium was isolated in Europe during 1789 and the element itself was extracted in 1842.

Uranium procuring in the United States was born in an environment known as the "Front Range" located, in part, forty-five miles west of Denver, Colorado. In 1871 Dr. Richard Pearce discovered a black mass of material on a dump resulting from gold operations. After having the mass analyzed it was discovered to be uraninite!

Since Dr. Pearce's discovery, a vast portion of all uranium mined in the United States has come from a spectacular and haunting area known throughout the mining world as the "Four Corners." This environment includes the northwestern thrust of New Mexico, the northeastern tip of Arizona, the southwestern portion of Colorado, and the southeastern quadrant of Utah.

Because of the ruggedness, age, isolation, and intensity of this region, uranium and associated properties have always been available primarily to persons of tremendous vision and strength. Those willing to chase radioactive ore bodies below the earth's crust often many many miles from domestic water in the most isolated and friable regions of the United States.

Substantial growth within America's uranium industry can be divided into "eras." The first, the "radium" era," occurred with the isolation of radium from uranium ore by the Curies in 1898. This period lasted from 1898 until 1924.

Vanadium, a property of uranium ore used for steel alloying, captured the world's interest and uranium was sporadically mined for this vanadium content primarily from 1916 until the creation of the Atomic Bombs rekindled interest in U_3O_8 itself. With the detonation of those bombs, the "uranium era" suddenly blossomed, then withered in the late 1960s. However, this century has issued new life into nuclear growth and uranium enrichment.

The foundation of our fledgling Nuclear Age is not the mass destruction of peoples; rather, visionary pioneers who dreamed of lighting continents and curing cancerous growth with the energy that comes from within the halo, halos of uranium ore.

Following the faint trail of pioneer industrialists, upon prehistoric earths of significant industrial and paleoevolutionary archaeological import, to glean secrets of our future as it pertains to domestic energies processed from "rare earths" and much humanity therein, as night descends upon New Mexico's Mount Taylor, the words of one who has trod this Brigham Tea Trail, Lawrence Clark Powell:

Martin Image "Indian Canyon Aspens" 1976

"One winter morning at Los Alamos the forest lay under snow, the people under covers, while steam from the laboratories indicated that the Atom was still monarch. At the turn-off to the National Monument in Frijoles Canyon, I recalled-Bandelier's THE DELIGHT MAKERS, the first novel of the Pueblo Indians who once inhabited the Pajarito Plateau.

. . . On these Southwest journeys, I am guarded from evil by the Sacred Mountains—Baboquivari deep in the Papagueria, Sierra Blanca beloved of Gene Rhodes, and Mt. Taylor and the San Francisco Peaks at either end of Navajo-land. West from Gallup the eyes seek first sight of the Flagstaff mountains when suddenly they appear on the horizon (like white sheep in a pasture, Haniel Long said) and the heart quickens. Only a hundred miles to go. What was it like in the time of the first white travelers when these volcanic cones did not afford the security of familiar landmarks?"

Naturally, the sacred earths to nurture nuclear potential's evolution tell their own story—a story which includes significant events in America's democratic and industrial growth since 1859, a story dealing with the recent past, and a story of the not too future, future.

As cold winds from the north and Canada inch into my weathered edifice erected from lodge pole pine and aspen saplings set against radioactive seven-thousand-foot high Temple Mountain, Utah just 25 miles north and west of Robber's Roost, Utah and on Robert Redford's "Outlaw Trail," my mind's eye perceives the vision of Lassiter or Annixter upon the horizon. One, well-dressed, single figure etched against ripple-marked shale and pre-historic Pinyon Pine many miles from domestic water—twenty, if you date drink the Dirty Devil River!

Lawrence Migliaccio, Vanadium King No. 6 Mine, Temple Mountain, Utah,

Some forty miles west of Denver an American adventurer, John H. Gregory, made slow passage up the north branch of Clear Creek which issues forth from the eastern slope of the Colorado Rockies. It was the winter of 1859 which found Gregory methodically prospecting this region. Only a lure of precious gold and other minerals which "gleam in the darkness" would bring a human being to the frozen Colorado Rockies in the depths and clutches of winter's fury. But this migratory Georgian had come West seeking fortune and winter just slowed his pace a bit. Little record remains of this man's sojourn yet it is recorded, with the thaw of winter's hold, when the rivers and creeks of the eastern slope of the Rockies flow with the most intensity and life unfolds virgin and green, Gregory was rewarded with the discovery of a semi-rich vein of quartz from which he mined and panned over $1,000.00 worth of that precious mineral, gold. Not being prone to lighting in any one area too long, Gregory sold his "claim" for some $21,000.00 and soon vanished from prevailing history. But his diggings, name, and remarkable feat of prospecting during the winter live on. Primarily, because several notable frontier journalists, including Horace Greeley, were upon those same mountains and soon reported Gregory's small el-dorado. Such reporting was directly responsible for a tremendous influx of mining and prospecting upon the eastern slopes of the Colorado Rockies the latter half of 1859 and into the 1860's, legend today.

"Gregory's Diggings" spread throughout the surrounding gulches, valleys, and mountains, eventually evolving into such hamlets as Silver Plume, Blackhawk, Nevadaville, and Mountain City, Colorado. Mountain City was to become the most well-known of these three, its name evolved into "Central City".

One must note that from its inception Central City chose to bear no resemblance to bear no resemblance to archetype mining camps and cities of American western history. It must have been ordained from somewhere within the geologic cosmos that the environs encircling Central City would be the "place of beginning" for the discovery of uranium upon the Colorado Plateau.

Not satisfied with the mere discovery of wealth, there were those of Central City who felt it essential that the chronicling of knowledge must accompany mankind's thrust upon and into the earth. As early as 1861 a MINERS' and MECHANICS' INSTITUTE was conceived in Central City and soon gleaned international respect. Few Eastern or European scientists studying the West ever failed to investigate this environment. Men of knowledge are always drawn towards "essence" and the atom has proven itself more essential than gold . . . yet the gleam normally captures first attentions.

To Central City came one easterner ordained to influence this entire region and, therefore, the future of mankind. Nathaniel Peter Hill was a visionary, metallurgical chemistry instructor at Brown University in Providence, Rhode Island, far from the Rockies and deserts to the west. When offered the opportunity to travel the eastern slope of these mountains for a Boston syndicate concerned with settling in Colorado, just fifteen years after the Mormons had concluded their historic Hegira and founded Utah, Hill leaped at such a proposition and was soon a frequent visitor to this portion of the West making several sojourns in 1864 and the year following.

These travels finally brought Hill to Central City. Upon his arrival he found the mining industry associated with the Nevadaville and Blackhawk region suffering from the first of many periods of recession, primarily, due to the fact that placer mining and other surface methods of procuring ore had used up the earth's quota of surface ore. The existing technologies all applied to placer mining and, with the absence of small surface metals, lay practically dormant. Mother Earth had afforded mankind five years of daylight in which to work and procure fortunes. By 1863 she was to demand more homage of the miner and seeker of precious "stuffs". The time had come for the miner to enter the very earth itself in search of riches. The time for hard-rock mining had come! Drawn by surface glitter and success associated with "Gregory's Diggings," the immediate environs attracted professionals from all segments of the world's mining industry of the 1860's. Paramount among those attracted to Central City were the process pioneers who marched through the industry applying such historically significant processes as the Crosby and Mason and the James B. Lyons and Company desulphurizer. Once again, the Central City earth had proved mankind's technologies to be infantile when concerned with extracting nature's essence. When placer turned to hard-rock mining as the 1860's slipped away, the miners of the Central City region were found to be making little progress in the quest of mineral wealth. Apparently the metallurgical instructor from Brown University had been sent West for a purpose unbeknownst to him, yet to become much clearer many, many thousands of miles away in a foreign country.

Returning East following his previous visit to Central City, Hill sailed for Europe and while in Switzerland made two essential acquaintances. First, he met Herman Beeger while visiting Freiburg. Beeger was then known as one of Europe's finest metallurgists and quite taken by the young man from Rhode Island and the information he shared on the mineral wealth of Colorado. Hill was equally taken by the vast amount of knowledge Beeger possessed on processing ores. While visiting Beeger, Hill was also introduced to one Richard Pearce, a young professor from Cornwall and, interestingly enough, associated with the few regions in the world known to possess deposits of pitchblende uranium. This, however, did not concern Hill, though that meeting was to affect the history of mankind's study of the atom's essence.

Written history does not say for certain the reason for Nathaniel Hill's journey to Europe but suffice it to say that while readying for his return to America he persuaded Herman Beeger to accompany him to Boston where both men promoted a concern called the Boston and Colorado Smelter Company. With this man's great vision, nurtured by world travel and continental United States travel and being an Easterner by birth, Hill had little difficulty in procuring the first capital for his and Beeger's smelting venture in Colorado. Eastern thought had been primed for the past 20 years with tales of the mineral wealth which America's West afforded and finding investors proved easy.

Returning to Blackhawk in 1868, with Herman Beeger's assistance, Hill set up his smelter. The miners were ready for new technologies and the initial thrust of this venture proved fruitful for all. There are those who have reported that this effort of Nathaniel Hill's saved the mining industry of Gilpin and Clear Counties, Colorado. Hill's smelter did not prove to be an overwhelming financial success because of the vast distances his matte had to be shipped for refining. Yet it afforded the industry of this rugged environment on the eastern slope of the Colorado Rockies the opportunity to endure long enough for one other visitor to reach the mountains from a distant land and another vision proved essential to the development of America's uranium industry.

Moab uranium men, Howard Wilson Balsley 3rd. from right.

While visiting Freiburg and making the acquaintance of Herman Beeger, Nathaniel Hill met Dr. Richard Pearce. Pearce, born in 1837 at Barrippa, near Camborne, Cornwall, had been associated with the mining industry of Europe since birth as his father was a mine superintendent by profession. Admirably schooled in Europe, Pearce always seemed to be learning in the very few environments on that continent where uranium was located. At a very young age Dr. Pearce studied with Europe's leading metallurgist. Naturally, when he met Nathaniel Hill at Beeger's in Freiburg, these two young visionary men were quite taken with each other. Young men of vision who admitted having the desire to probe and study the "nature of all things" though keenly aware of the infancy concerning man's study of his earth, in the 19th century, were kindred souls!

Whether it was more than mere "business" that drew Dr. Pearce to Central City, Colorado, in 1871 will never be known. Yet great men are normally drawn to the "essence" of most matters for reasons which cannot be chronicled. He came to Colorado to study a group of mines owned by the Rochdale Mining Company.[4] His close relationship with Hill and Beeger was renewed far from Freiburg.

One of the mines Dr. Pearce was to inspect had a great history of producing gold. This was the famous "Wood" mine well documented in the history of our American West. Since the thrust of humanity upon this eastern slope of the Colorado Rockies had been centered on the mining of gold and silver, naturally, any other material which carried no trace of those precious ores was discarded. In fact, the miners of Gilpin County had for some time been annoyed by masses of dull black material which showed no traces of gold but because of its consistency, often caused the smelting process to break down and was generally a nuisance to the miners.

While exploring a dump associated with the Wood mine, Dr. Pearce was to make a discovery which, most certainly, has affected Western thought and deed. He was attracted to one such mass of black material and his pulse quickened for this man of science had seen substances much the same as this mass in Europe. At a feverish pitch he collected samples, mounted his horse and traversed the trail to Blackhawk where he knew his mentor Beeger had the apparatus necessary for analyzing the black substance. As the pines rushed past on both sides of the trail and the crispness of the Rockies pressed against his person, he knew the sojourn to America had been for a reason. Yet being of scientific mind he understood tests must be made.

Neither scientist could conceal their excitement and expectancy as they worked through the day and into evenings blackness. With nights descent upon the Rockies of the 1870's most slumbered with relish. As each second of history passed and those western heavens descended upon the laboratory, a strange stillness must have permeated Blackhawk. For these two men were discerning a force eternal in all of nature.

The keen vision of Dr Pearce and Herman Beeger's "method" was soon rewarded. Tests proved positive and the black mass was discovered to be uraninite! Both scientists and frontiersmen, these two were overwhelmed with the discovery. Never before had Pearce seen uranium in such quantity, and then only in very certain European environments. Strange that blackness is not always void.

Immediately thereafter, Pearce returned to Europe but not before having some several hundred pounds of the mineral sorted and made ready to ship to him. Upon its arrival in Europe, he ran further tests and ended up selling it to Johnson and Matthey of London to be used for experimental purposes and especially for dyes, inks and stained glass.

Returning a year later, he came to the eastern slope not as explorer and scientist, rather as a miner intent on mining the Wood for its uranium content, not gold or silver. Able to procure the lease, he was moderately successful at mining the ore already exposed. Naturally, mining virgin ore, that which he was able to collect assayed a very high percentage of U_3O_8. Dr. Pearce did not, however, become independently wealthy mining the Wood and soon returned to activities less physically exhausting. But he had paved the way for this industry new to the United States to grow. Rapid progress considering it was only 12 years since John H. Gregory first prospected the Clear Creek in the winter of 1859.

Martin Image, young, Ute Tribal Lands, Utah, 1976

This fledgling industry then suffered from much the same malady as it does today. Yet the century between 1871 and 2012 has seen this black mass used to destroy a peoples with its force, cure cancer with the same force, and light up cities with the same force. One problem then as now, the proper refining process. How does mankind refine the eternal atom?

Not one to let the genius of Pearce go without challenge, Nathaniel Hill offered him the opportunity to head up the Boston and Colorado Smelter Company's Blackhawk mill. All the while, Dr. Pearce had been making his discovery and mining the uraninite, Hill had been fast at work constructing another smelter in Park County, Colorado. He was already fully aware that without the proper technologies for processing the crude ore what little early profit did exist was eaten up with transportation and shipping costs.

It cannot be denied that Dr. Richard Pearce did discover the first deposit of uranium in the Western United States. His place in the history of atomic development is secured. Not only was he paramount in the discovering of the mineral in Colorado, but was also the first to hand mine such. Though his efforts at physically procuring the ore did not last very long, it was accomplished. Not long thereafter, he was appointed Her Majesty, Queen Victoria's vice consul in Denver.

Hill was finding it difficult to make money on his operation yet resolutely continued. This was a shrewd pioneer and when the smelter at Blackhawk was not able to pay its own way in the fledgling enterprise, he began the construction of yet another smelter at Argo, Colorado, a little closer to then prevalent modes of commerce. Upon completion, he was forced to let the operation at Blackhawk cease in 1878. However, uranium in America was no longer Mother Earth's secret.

The earliest geological report mentioning the ore deposits of Central City and environs appeared in a 1913 issue of the MINING AND SCIENTIFIC PRESS published by T. A. Richard:

The Kirk, German and Wood (mines), all on Quartz Hill, are the most important and best known. They lie in a north-south belt less than 1,000 feet wide and perhaps half a mile long. This is the area where all ores in any way identified with the production of pitchblends are found.

In this Central City district, gneiss and crystalline schist predominate. In association with the veins of these mines there occurs, in intrusive form, a fine-grained aplitic granite closely associated with the pitchblends. This granite takes dike form, sometimes tongues, more rarely connecting sills.

The veins themselves without exception, have a well-defined northeast-southwest trend. Similarly, without exception, the fine-grained granite intrusive is common to all (veins) sharing in the pitchblends distribution. It is significant that some intervening veins, yielding profitable gold and silver ores, but not pitchblends, are bounded by walls of crystalline metamorphic schist, the granite intrusive being absent. This plainly suggests the derivation of the older fine-grained granite from a main mass, which in its intrusion into the fissures in the state of magna, favored some and not others of these veins.

In 1952, B. Moore and C. R. Butler continued this thought in the GEOLOGICAL SURVEY BULLETIN:

"A series of faults that strike east to north east and dip steeply north or south cut all the rocks on Quartz Hill."

It is therefore reasoned that these faults acted as flumes for hydrothermal liquids which carried and then deposited the urantite.[6]

The time has come to descend the mountain and move westward to the intense grandeur of southwestern Colorado. Down the slope of the Rockies this chronicle cascades to rest in the cattle country of western Montrose County. For the "essence" of uranium was not held captive high in the Rockies alone. The paradoxical contours of the "slick rock" country; gentle inclines leading to treacherous cliffs and swirling waters warn man that nature on this lower plain can be intense.

With Dr. Richard Pearce hundreds of miles away far below the eastern summit of the Colorado Rockies, two brothers of Montrose County, called the Talbots, came upon a rather remarkable outcrop of ore while herding cattle. The location was on Roc Creek, a swiftly flowing stream of clear water about 15 miles north of the frontier town of Paradox. Having no idea what this ore might be, but hoping it to be silver, they sent a sample to the American Smelting and Refining Company's smelter located at Leadville, Colorado. A town known today more for its world famous mining museum To their disappointment, the assay report explained that the smelter had no idea what this ore might be, but silver it was not! With this negative report, they decided to stake no claim on the discovery. The year was 1879.

Little was done in regard to locating new deposits of uranium in the United States for the next fifteen years. A small amount was procured from the Central City District, but little development occurred further west. How many frontiersmen, such as the Talbots, witnessed strange ores outcropping from the Salt Wash or Mossback that later turned out to be uranium is unknown. Basque sheepherders, cattlemen, gold prospectors, timber men, railroad men, and American Indians all traversed the "four corners" as the 20th century approached, yet the intensity within uranium ore lay strangely dormant upon the uplift of the Colorado Plateau. The Quaken Asps danced, then slept, danced then slept, as summer turned to autumn, autumn to winter for many, many years. Yet below the Plateau and upon the San Rafael Swell halos of uranium ore shown in the darkness.

Uranium was not to slumber until the 20th century however. It is discovered that an old Irishman had staked a claim on this same fissure that the Talbot had uncovered some 16 years before. The Irishman, Tom Dullan by name, had filed the claim in 1895. He also had no idea what the strange substance was and had turned several samples over to a Frenchman named Poulet. Poulet, a young graduate of the PARIS SCHOOL OF MINES, was then doing assay assignments at the Cashin copper mine not far from Paradox, Colorado.

Unable to make the identification for Dullan, Poulet had the samples sent to the PARIS SCHOOL OF MINES. Filed in the archives of the PARIS SCHOOL OF MINES is the report which found Dullan's sample to contain uranium!

Unlike the black mass pitchblends of the Wood mine on Quartz Hill, this new ore displayed remarkable yellow shades for it was found as an outcrop and being exposed millions of years to the southwestern Colorado climatic extremes had given rise to oxidation processes. It was from this yellow ore that the second major type of uranium mineral was named. That being carnotite in reference to the great French physicist, Adolphe Carnot.

Howard Wilson Balsley, ore truck breakdown.

TEMPLE MOUNTAIN, UTAH

Tom Dullan's claim passed through many hands for the next several years. It finally was procured by one Jack Manning, colorful frontiersman and legend upon the Colorado Plateau. Manning was the first to name this claim and the outcrop along Roc Creek was to become the famous "Rajah." The first mine so named because of its uranium content in America. Because of the treacherous terrain in which "The Rajah" was located, little development work was done on it by Manning. As today, the vast portions of the Colorado Plateau were virtually inaccessible in the 1890s. All ore was hand drilled or picked, then transported out of the canyons by mule. This practice carried into the 20th century and as late as 1941 Lawrence Migliaccio of Price, Utah and John Adams of Green River, Utah, hand-drilled many tons of high-grade uranium ore from near the peak of 6773 foot Temple Mountain in Emery County, Utah, and packed the ore down the mountain on their backs to mules below. It is reported that this high grade ore was used in the making of the Atomic Bombs which helped end the Second World War. At the time neither Migliaccio nor Adams had any idea what the result of their efforts would be.

But history is the fact that humanity has often virtually clung to the breast of Mother Earth, though her whisper could send one tumbling into eternity in search of that "gleam." And those persons involved in the quest of the halo called "uranium" have always clung precariously to the breast of their earth yet they have clung!

Pete and Hattie Dalton Shumway of Moab, Utah, reported using an improvised aerial tramway powered by the jacked-up wheels of their Model-T Ford to lower ore from the heights of Bridger Jack Mesa in San Juan County, Utah, to the sage and desert below before 1940. A great southern Utah mining family, the Shumways have often been alone upon the deserts of the Colorado Plateau in search of the atom, one rock at a time.

The desert blossoms, the slick rock glistens in the afternoon sun following a warm spring shower from Mexico whose journey abruptly ends at the western slope of the Colorado Rockies a little to the east of Dolores, as this search for uranium approaches our 20th century.

By 1898 "The Rajah" lease had been procured from Manning by Gordon Kimball of Ouray, Colorado, who had learned of the carnotite associated with the Roc Creek area from Poulot whom he had met in Denver. For the first time in the history of the uranium industry, a party was interested in the ore for purely business reasons. An industry was truly born in June of 1898 as Kimball hand mined ten tons of the yellow ore, and was eventually paid $2,000.00 for this act.[9] Thus, in the twilight of the 19th century, a handful of human beings had been drawn by intensity to the plateau called the Colorado and set upon that earth at great peril to their very persons, respecting the grandeur and subtles of this remarkable environment. They had learned one must travel a great distance finding the quietitude essential for listening uninterrupted to their earth's narrative. Thus had not come to destroy, rather, search for life using but their bare hands and keen vision. And the earth did not object.

However, with Marie and Pierre Curie isolating radium in France... to cure cancerous growth... two other Frenchmen already chaptered in this chronicle were working for Madame Curie in Colorado's slick rock country. They were assembling the very first uranium concentrating plant in the world, on United States soil, upon the banks of the Dolores River at Camp Snyder in San Miguel County, Colorado, near where Colorado State Highway No. 80 now crosses the river.[10] Mrs. Nellie Snyder, a pioneer woman of the slick rock, has reported the mill being erected on her property in 1898, and allegedly shortly thereafter Madame Curie visited the site and is said to have sojourned the southwestern portion of Colorado rather completely in 1899. As well as over to Temple Mountain on Utah San Rafael Swell.

With the slow turn to the 20th century this chronicle, of course, looks a little more to the western expanse for the answer may follow the rivers and those rivers all flow west, west of the Continental Divide. Does not that massive river of life, the Colorado, flow through Colorado and into Utah? So follows mankind's search for uranium.

By 1900 the number of persons seeking uranium ore in the western United States had appreciated noticeably. Yet "industry" was still founded on rugged human beings challenging a frontier one rock at a time, relying on fortune more than all else!

To the west of Colorado's Dolores River in Utah, a pioneer from Moab, by the name of Turner, located the first uranium claim on record in the state, the year was 1896. This act appears to be visionary formality and little else. Such scant knowledge of the mineral existed at the turn to this century that each observation recorded pertaining to carnotite and pitchblende was significant document. Man had not yet entered the earth to any extreme in quest of this energy.

By 1910, names such as Gregory, Bill, Pearce, Beeger, Dullan, Poulet, and Talbot are not as often uttered within circles of those seeking both the substance and essence of uranium. Rather, there appeared many more efforts to challenge the isolation and extremes of the Colorado Plateau and San Rafael Swell by groups of individuals banded together in the quest of uranium. With the Curie's internationally acclaimed work in France concerning the isolation of radium from uranium ore, companies such as the Standard Chemical Co., Rare Metal Co., and Dolores Refining Company all had crude refining facilities in western Colorado. The influence of these companies reached far into the Utah and Arizona frontier. Persons who sojourned the Slick Rock and deserts of southern Utah did not travel quite so alone as once man had. Perhaps this is a bit more figurative than literal for all who set upon this earth have traveled alone. By 1910, the "Corporation" had come to the sage!

Martin Image, Lester Chapoose, Chief of the Uinta and Ouray Tribal Lands, Utah, 1976

What better environment is there to recount the birth of corporate thought during the early 1900s yet trace the faint trail of prototype human beings concerned with mankind's search for the essence of life, than upon the Great Domal Uplift of Utah's San Rafael Swell? (Where earths were painted by Klee and God.) Through which the largest reef in the world passes then disappears into the Goblin Valley. For few who traversed the slick rock or red cliffs of Moab dared to set upon the intensity of the Swell as two centuries of American history passed each other in the dusk.

A frontiersman by the name of Swazy is reported to have seen traces of yellow oxidized uranium ore upon the San Rafael Swell in the year 1893 near the most prominent up-thrust upon that Swell, 6, 773 foot Temple Mountain. (There are two paramount Temples in the territory of Utah. One rises above the Great Salt Lake Valley, while the other raises above the great Southern Utah Desert.)

As the Talbot's had before on Roc Creek, Swazy did little other than make a mental note of his discovery. The heat of the deserts' sun makes too much exposure to the San Rafael Swell impossible. Most have simply attempted to complete their task then abandon the Great Domal Uplift. Even today, a great silence permeates the washes within the "Temple's" dominion. So it was at the turn to this century.

In 1904, an American-born geologist, J. M. Boutwell, was told that "considerable deposits of a black, vanadiferous sandstone and some carnotite float" had been discovered in "Wildhorse" Canyon in 1903. Sheepherders, too often the butt of abuse and scorn within the written history of the American West had been searching for what moisture and life they could find southwest of Temple Mountain and had made the discovery. Boutwell reports the find was "later prospected in dense, carbonaceous sandstone which contained, remarkably enough, combustible matter." When that process occurred, the remaining residue was found to contain traces of uranium by-product, vanadium! Boutwell admits to never witnessing these workings in person. But his keen understanding of the San Rafael Swell is essential in placing this discovery:

Wild Horse Creek empties into the Muddy River about 15 miles southwest of Temple Mountain. Upstream from the junction the creek passes between the Morrison capped East and Middle Wild Horse Mesas and then skirts the eastern escarpment of the San Rafael Reef. It splits about 2 miles south of Temple Mountain, with the left form turning west and cutting through the reef as Wild Horse Creek. The other split continues an additional mile before it enters the reef as South Temple Wash. From Boutwell's brief description of the ore it is believed that the discoveries he mentions were in the Temple Mountain area, for it is conceivable that the split now called South Temple Wash could have been known earlier as Wild Horse Canyon. The numerous old workings along the outcrop of the Chinle Formation in South Temple Wash indicate that the initial discoveries of oxidized ores were easily accessible from the Canyon.

By 1914, the lure of Temple Mountain's colored earths proved too intense for man to ignore. Because of Temple Mountain's isolated nature, little record remains of man's activity just west of the largest reef in the world before 1920. Yet adventurers and frontiersmen reported small companies of men being sighted upon the Swell intensely probing the earth only to vanish from sight a short time thereafter. One such visionary, F. L. Hess, visited Temple Mountain's environs in 1914. He had traveled from that oasis upon the banks of the River, on the trail to Grand Junction, Colorado, by horseback to study and observe what he found to the west of the reef. No permanent trails existed to the south of Green River; therefore, each mile was virgin to the traveler upon the San Rafael Desert in 1914. Early wayfarers to this desert invariably pressed as close to the reef as possible while journeying south towards the Goblin Valley.

Martin Image, portrait Ute Tribal Lands, Utah, 1976

Roughly forty miles southwest of Green River Hess spied the "tower-like" peaks of Temple Mountain rising above the 2000-foot sheer face of the San Rafael Reef. This signaled the first of many times Hess was drawn to Temple Mountain and for the next five years he was to be a frequent traveler to the environs west of the great reef, that land in which a foot of snow can fall in one hour after a day of 80-90° temperatures. But his findings of 1914 prove the most essential to the chronicling of man's search for energy. He reported that during 1914, the "commercial production of uvanite, a radium-bearing mineral new to science, accompanied by other uranium minerals, one or more of which are yet to be described, was begun at Temple Rock, 45 miles southwest of Green River."

Though Madame Curie had isolated radium by this date and several notable companies had set up uranium processing facilities upon the

Colorado Plateau, men were mining radioactive ores from Temple Mountain which had yet to be named and "distinguished" by science! Perhaps the time has now come in the written history of man to look a lot closer at the growth of atomic potentials. For it is still not recorded what potentials were found just west of the reef in the year 1914. All that has been reported is the presence of Japanese miners, because their size afforded them a greater mobility when entering the North Face of Temple Mountain at a point 2,500 feet above the desert below. Yet they were spotted pressed against the mountain in search of essence during 1914. The story of mankind's study of the atom has normally unfolded within environments far beyond the visions of most therefore shrouded in secrecy.

While "unnamed" radioactive properties were being procured from Temple Mountain, the United States Government was becoming involved in man's search for essence. On October 7, 1913, a historical joint venture into man's search for the atom was initiated by the American Scientific Society, Government, Medical and Business societies. The resulting effort was the National Radium Institute. Its purpose was to develop America's radium potentials and share this effort with the world. The head of the Institute was Dr. Howard A. Kelly, a John Hopkins University cancer specialist. With the aid of the Bureau of Mines, Kelly had obtained leases on 16 carnotite claims in western Montrose County, Colorado.[13] Ore from these claims was to be prepared for shipment to either of two processing mills then in construction. One mill was being built south of Denver and the other on the San Miguel River near Naturita, Colorado. The Institute was, however, solely interested in the radium content of the uranium ore. By 1917, the National Radium Institute had obtained enough radium for unlimited application in the medical and scientific world of its intention, and was disbanded. Thus ended a very short, yet humanitarian enterprise within our American experience. America's first efforts at understanding the potential of the atom were directed towards retarding and destroying cancer. For those of Temple Mountain have often made light where there was once only darkness!

Upon Emery County Utah's San Rafael Swell, miners had continued to slowly inch their progress up the North and South Temple Washes to the slopes of great prominence. No record remains of mining development during 1915 and 1916 upon the isolated and eroded terrain of Temple's Moenkopi "Sinbad Country." However, once again F. L. Hess reports traveling the desert in 1917 and observing mining activity. In 1917, the United States Radium Corporation's auxiliary concern, the Chemical Product's Company (whose chief chemist and physicist, Dr. Howard H. Barker, was soon to gain international attention for his work with Dr. Hermann S. Schlundt of the University of Missouri in the development of a standard technique for determining radioactive fixation in a living organism) was fast at work mining the uvanite ores from Temple Mountain. Later in the year they were to have radium salts ready for crystallization. Hess noted, "in the area being mined the conglomerate was approximately sixty feet thick and cross-bedded, the carnotite bearing asphaltite being mixed with the predominating coarse sand." Amazingly enough, the mountain abounded in "lignitized, petrified tree trunks," indicating that this barren environment was once a great inland forest.

1918 was much the same as the previous year. The Chemical Products Company of Denver was continuing to isolate radium from the uvanite ores obtained from "Temple Rock" and then bought the claims.[15] Mining upon this great mountain was slow in 1918. Domestic water had to be hauled in wagons drawn by four and six-horse teams across a land where roads were non-existent. Yet the lure of uranium's essence continued to draw learned miners to the Sinbad and beyond! What little moisture was naturally associated with this environment was naturally associated with this environment was found in potholes high upon the San Rafael Reef. These eroded holes in the slick rock were later called "Mormon Bathtubs." A subtle tribute to the great Hegira of human sacrifice accomplished by Brigham Young in 1847.

The intensity of extremes associated with the great southern Utah deserts proved too taxing for Dr. Barker and the Chemical Product's Company of Denver, and, the claims located within this once great forest, now petrified, were sold to the Ore Products Corporation in 1919.

Little did those miners and prospectors of the Colorado Plateau whose existence was predicated on the search for uranium ore know that a dark cloud slowly sojourning the Atlantic Ocean was to settle over their "four corners," in a short time, and this earth would be asked to cease its efforts in regards to isolating the atom for the cure of cancer.

With the dissolution of the National Radium Institute in 1917, American uranium ore producers found that the market for radium in the United States had vanished. It had taken scientists of great vision to form that Institute, in the first place, and those persons were weary yet possessed vision to procure enough radium to fill their every need. Therefore, the independent operators who had relied on the Institute's mills south of Denver and on the San Miguel River in Montrose County, to buy their ore were forced to look for foreign influence and markets to sell their bounty. Europe was, once again, the greatest market for American effort.

As cold winds from Canada moved across the Plateau casting shadows of life and death upon the earths below, the 1920s were issued into the American Experience. While true frontiersmen of this decade still eat huddled around fires of desert sage to cast off January's chill, the uranium ore they were cutting away from the earth below, that ore which "contained combustible matter," (matter which latter killed Madame Marie Sklodowska Curie) had not yet been equated with domestic energy by the miners themselves. Their ore was being used by foreign principles, and a pinch was used in the United States for inks and dyes; human beings of America circa 1920 were wearing this radioactivity on their bodies, day in and out, unbeknownst to them. But this was the world of the procurers of uranium some 80+ years ago.

Slowly, that dark cloud descended upon this handful of pioneers who prospected and mined the Swell and slick rock.

Martin Image, portrait, Ute Tribal Lands, Utah, 1976

Far across the Atlantic Ocean the largest discovery of Pitchblende uranium in history was made in the Belgian Congo, an environment as extreme as the Temple's "Sinbad," an environment closer to the radium markets of Europe. Also, in 1922, the western United States' earth rested as the need for American radium vanished. Desert sands and sagebrush piled up in front of the abandoned Temple Mountain Mines. A strange quietitude settled upon Colorado's Paradox Valley. But keen minds were not stilled and would not be forced to slumber.

The fledgling uranium industry of America did not slumber for long under the chill heavens from Canada. Another property of uranium was soon to serve as the life saving influence upon this industry. At roughly the same time prospectors, including Jack Manning and Gordon Kimball, were taking notice of uranium ore upon the Colorado Plateau in 1898 (with the exception of Dr. Pearce and Poulet's work in the Central City Mining District). French chemists were studying the amazing properties of another element contained within the mass of uranium ore. That property was vanadium. They found if ferrovanadium (a compound of iron and vanadium oxide) was applied to vats of molten steel, the tensile strength and elastic limit of the steel would increase greatly.[14]

Earliest record of any knowledge concerning vanadiums existence in ores mined from the Colorado Plateau was made at the turn to this century by J. M. Boutwell who is also remembered as the geologist first responsible for reporting the existence of uranium ore in the Temple Mountain area. It was he who told of the sheepherders' 1903 discovery in Wild Horse Canyon. Aware of the experimental work being conducted in France, he reported:

"The value of vanadium and uranium for commercial uses is stimulating a search for compounds of these rare elements. Vanadium which is used chiefly for hardening steel (it is claimed to be twelve times more effective for this purpose than tungsten) is scarce and this utility and scarcity tend to create an increasing demand."

While industries' attention had been focused on radium until 1922, a notable quantity of this vanadium was mined for military purposes. Dr. Gary Shumway has reported that by 1912 over 600,000 pounds of refined vanadium was produced on the Colorado Plateau.[16] This quantity grew to over one million pounds by 1917 as American mines were producing at a feverish pitch to supply the Allied Powers with this precious mineral essential to adding strength to the steel plate used on tanks and battleships. Therefore, once again the essential role of uranium was in the "protection of human life."

The 1922 discovery in the Belgian Congo did, as stated before, force the cessation of mining for both the radium and uranium content of uranium ore. The atom had proved too elusive to nurture further interest in eternal energy. In a sense, Mother Earth decided she had shared enough with western thrust; for the lonely quaken Asps knew of the atoms eternality yet mankind had not gleaned this essence. In her beneficence she allowed, again, a very select few of vision to develop the tensile strength of steel so that association with uranium would not altogether cease. And then, often persons possessing traces of mystical quality were allowed to associate with uranium as the Great Depression of the late 1920s was being formed, much the same as uranium, below surface tranquility in halos of intensity slowly pulsating towards fresh air!

Martin Image, John LaRosa, Ute Tribal Lands, Utah, 1976

ONE SUCH VISIONARY

One such visionary possessing traces of mystical quality, left the humid plain upon which the City of Indianapolis was founded and began a slow sojourn across the plains to Denver then traversed the

Rockies to rest within the red cliffs of Moab, Utah. The legendary Howard Wilson Balsley arrived in Moab on his birthday, December 7, 1908. He had come west as a speculator for he had invested money in an irrigation project that was being promoted in Indianapolis. The project was unfolding near Cresent Junction, Utah, and Balsley felt bringing water to the deserts of Utah, if it proved successful, would benefit mankind and be a fruitful business venture. The desert never blossomed and Howard Balsley was left penniless in Moab. Yet even schemes which fail when concerned with making the deserts of America fertile are worthy of note, especially if the year is 1908.

This is a person responsible for America uranium industry enduring those years after the Belgiun Congo strike silenced foreign need for American uranium products. At times, so alone upon this American frontier, Balsley's vision and endless work promoting uranium's potential may be of paramount importance in the history concerning atomic potentials.

How did his vision expand in the west? In 1909, he became a forest ranger, frontiersman in essence, and learned to listen for the lyric of nature. He reports that in 1917, he received orders from Washington to classify all of the fertile agricultural possibilities west of the La Sal National Forest, which at that time embraced the La Sal Mountains, the Blue Mountains of Monticello, the Elk Mountain north and south and the Shay Mountains. "I was assigned the task of mapping out all the agriculture lands inside of the national forest." And so, often on foot, he traversed this veritle shangrala. The earth must have long before sensed Howard Balsley's respect for all life. Few are still living who can accurately recount their hours spent upon the "four corners" before 1919. One year in this environment can be the experience of many lifetimes, for most.

Not only did Howard Balsley learn to converse with nature and communicate with Washington, D.C. from the distant glades of the La Sal National Forest and red wingate cliffs of Moab, Utah but the true native American was a part of his daily experience:

"NEW OLD MANCOS JIM was chief of the renegade Utes and he would come into my office to get what he called "Washington Paper Talk." He liked to impress the other Indians with legal papers, so I would give him a crossing paper or wood permit and show it around and say that it was "Washington Paper Talk." I once asked Mancos how many snows he was old and he held up both hands with fingers outstretched ten times and with one hand once. That would have made him 105 years old. I can believe it as my father-in-law was an old-time cowboy who had come to the Canyonlands at an early age and he said Mancos Jim was old then."[17]

And so, by 1923, Howard Wilson Balsley had come to the west and had come to know the west.

In 1923, the United States Radium Corporation of Orange, New Jersey, ordered four carloads of carnotite ore containing at least three percent uranium from the Yellow Circle Mining Company.[18] The head of this mining company was Howard Balsley. Balsley grubstaked a prospector who had dreamed of a "yellow circle" upon the desert indicating the presence of precious ore in the upper Cane Springs area of San Juan County, Utah in the spring of 1915. Balsley recalled that the sincerity of Charles Snell when telling of the vision proved irresistible. The same Snell soon discovered a remarkable group of claims in the Upper Cane Springs area with a very distinct yellow circle etched in a rock.

Balsley's contract with the New Jersey concern issued a new hope for the production of uranium ore by the independent miners of the Colorado Plateau, yet, by late in 1924, the boom had not taken place and the uranium industry of Utah became essentially one man for the next ten years. That man, of course, Howard Wilson Balsley.

Howard Wilson Balsley magazine photo.

From 1924 until early 1934, the market for uranium ore had bottomed out and what was bought by eastern concerns was, for the most part, mined from the Yellow Circle Property of Howard Balsley. The rich ore was hand drilled and hand sorted by Balsley and his crew for use in making luminous compounds and medicines. But this was a very small industry and the deserts of southern Utah had not let up in intensity yet Howard Balsley kept those who would set upon that desert in search of uranium in motion. Alone, upon the banks of the Colorado as it knifes through the red rock of Moab and slowly travels towards Dead Horse Point, Howard Balsley often pondered the future of man as the silent river of red earth passed his person. But he knew of uranium's potential and if he must sort the future of man with his bare hands, then sort he must!

As a point of fact, the United States Bureau of Mines reported that in 1932 there were only two parties who produced any uranium in the state of Utah that year. They were, Howard Balsley and two brothers from San Juan County, Utah, "The Shumways." That same year only three parties produced any uranium ore in the whole of Colorado. The essence of the atom was gathering intensity unmolested by human energies in the year 1932. But the visionaries are always there!

By late 1934, Marie Curie was dead, Poulet was dead, Dr. Richard Pearce was dead, Nathaniel Hill was dead, yet uranium lived!

As the spring of 1934 blossomed upon the slick rock and occasional rains washed massive Temple Mountain, to glisten as a 6773 foot silica diamond upon The Great Domal Uplift, Howard Balsley negotiated a contract with the Vitro Manufacturing Company of Pittsburgh, Pennsylvania that is of significant note within mankind's development of atomic reality. For this man who had learned to negotiate with the likes of "NEW OLD MANCOS JIM" had negotiated life into a dead industry. Had he not sojourned the trails from Blanding to Monticello to Moab, then Cisco, Thompson to Green Rover to Dolores and Naturita for 10 years, alone, the miners of vision in these same hamlets would have lost all vestige of desire to front the Colorado Plateau, also alone, in the quest of the energy from within the atom. Howard Wilson Balsley rarely was found to rest.

As the chill of a winter's night inches into the room which has served as a place of association with Howard Wilson Balsley for over 40 years, he recounts eleven years of life essential to your well being in five minutes, for he had already acknowledged that great vision is often but symbol:

"During the eleven years I was affiliated with Vitro, I was the only ore buyer on the Colorado Plateau who paid for both the uranium and vanadium in the same ore. There were a couple of other buyers who would pay for the uranium content only, so I got most of the ore that was available from the smaller producers. Of course, during most of that time, the United States Vanadium Corporation, a wholly-owned subsidiary of Union Carbide Corporation, was operating vanadium-extraction mills at both Uravan and Rifle, Colorado, and the Vanadium Corporation of America had a mill at Naturita; however, they only paid for the vanadium content of the ore."

Of essence? This industry lived for those eleven years of the American Experience. Union Carbide Corporation of America and Howard Wilson Balsley often face to face in the most desolate reaches of America west. Over eleven calendar years passed as these two forces were felt throughout the "four corners." Naked before the expanse which surrounded them they kept the pulse of the atom alive within their immediate experience. The gentle visionary from Moab proved to be a pillar of fortitude!

In 1940, the persons of America's uranium industry were to soon be called to duty. They had been waiting since the turn to this century upon the Colorado Plateau and San Rafael Swell for a greater national understanding, need and compassion towards the remarkable mineral which had lured them to the isolation of the many great deserts, for many years, of the quiet essential to true understanding. Visionaries such as Lawrence Migliaccio of Price, Utah and Howard Wilson Balsley, Jack Turner, Fendall A. Sitton, Raymond A. Bennett and a myriad more had been afforded many, many years of such silence to ponder this remarkable energy of intensity. Now as the United States entered the conflict of the Second World War, the need for uranium and its associated properties was evident, that need? To protect the democratic societies which had nurtured it for humanitarian purposes!

Late in May of 1942, the United States government, given two years to organize after being attacked by Imperial Forces, set up a corporation to procure as much vanadium as possible. That organization was titled the Metals Reserve Company. Those close to the industry on the Colorado Plateau were pressed to produce enough of the precious substance as the industry was still predicated on men of strength and vision caressing their earth one stick of dynamite at a time and hand picking and drilling the ore. Few technological advances in uranium mining had occurred since Kimball first mined "The Rajah" in 1898. And by 1942, the United States was fast at war! Young American men from unnamed hamlets in the hills of eastern Kentucky were falling to the soil in foreign lands never to breathe again while each minute of

American history ticked from the "westclocks" of Trenton. Yes, America was at war!

The government then set up several ore buying stations on the Plateau and was buying all vanadium. Then, as abruptly as the program for vanadium was set up, it was also closed. Late in February of 1944 the government decided that enough vanadium had been accumulated and those along the trails of the Colorado Plateau were suddenly without a source for their ore again![18]

Early in the spring of 1941, the vision of a man upon Utah's San Rafael Desert was discerned by the legendary Andy Denney. Any form, save buffalo and herds of wild horses, elk, mule, deer, coyotes and other assorted wild life, was reason for attention upon the San Rafael Swell whose only known inhabitants were Andrew Denney and his wife. They lived at the foot of the San Rafael Reef about 43 miles southwest from Green River, Utah, and one had to follow very faint trails of those who passed before to reach the Denney Ranch. Few vistas in the world could have provided as much majesty as the view from the porch of Denney's Ranch. Unless a blizzard had set upon the Great Southern Utah Desert, one could clearly discern the intrigue of Utah's Goblin Valley to the southwest, the well defined Zane Grey's wild horse contours and the Gilson Buttes to the south, and Robber's Roost (the barren waste upon which Butch Cassidy and the Sundance Kid along with the Wild Bunch hid from authority). Plainly visible from the ranch, the La Sal's to the east often framed Robber's Roost while a bit to the west the Henry Mountain chain were prominent. It was certainly a location of one with great vision.

Martin Image, Image from portal entrance of Vanadium King No. 1 Mine.

The image of the man Denney had observed that February morning traveled to the west past his ranch for roughly three miles, then turned north only to disappear into the San Rafael Reef. To the west of the reef the towering peaks of Temple Mountain caused Denney to shudder. Lawrence Migliaccio, a well known figure in the state of Utah by 1941 and often a visitor and "worshipper" of its silent reaches, paused with relish as he reached the imposing walls of North Temple Wash unaware that Denney had witnessed his progress. Migliaccio had come to the Swell in search of uranium whose essence had intrigued him from his youth. Little did he know that this sojourn, one which he had made before, was to pave the way for his becoming a spokesman for all miners who face their experience with compassion and respect.

Migliaccio had come to Emery County Utah's Sinbad Country to mine uranium, not Vanadium, and much research had revealed to him that Madame Curie had procured ore, through Poulet's efforts, from very near the peak of 6773 foot Temple Mountain at the turn to this century. It made perfect sense to Migliaccio to seek the essence closest to the sun's energy and on the Colorado Plateau, Temple Mountain stood as monument to this very fact. Migliaccio was not alone upon the desert for other young men of vision were seeking of this essence. But Migliaccio was the only one reaching for western skies in his quest. Only one at a time has ever stood upon the Kayenta Crown of Temple Mountain!

By July, Migliaccio and John Adams of Green River, Utah, had scaled the north face of Temple Mountain where they commenced to hand drill and pick over ten tons of high grade uranium ore. Day in and out they could be seen almost 6,000 feet above the desert's floor picking away at the essence closest to the sun. Each rock at a time was carefully examined then the men would place up to 100 pounds at a time in canvas sacks and placing the sacks upon their backs, descend the mountain to the mules below.

The location of their diggings was well over fifty miles from domestic water, so Migliaccio and Adams learned to seek the shimmer of the quaken asp whose gentle dance normally signified traces of water. If none were to be found, they climbed high upon the Reef seeking pockets of moisture held within the Navajo Sandstone structure of that bioherm.

With the ten tons of ore secured to their mules, the men herded them out North Temple Wash to a battered truck whose duty it now was to carry this essence across the shifting deserts of southern Utah and deliver it some several hundred miles away to the United States Vanadium Corporation processing mill located in Naturita, Colorado.

After delivering the ore to Naturita the men returned to Temple Mountain and Lawrence Migliaccio continued to prospect upon the great mountain until the snows of December 1941 drove him from the Swell and forced his return to Price, Utah, some 100 miles to the north.

Denney later reported, from 1940 until 1944 Lawrence Migliaccio was the only person to frequent Temple Mountain and procure ore on a regular basis. During these years he was alone upon the Swell, Migliaccio, learned many of the peculiarities of this isolated realm. He traced the paths of humanity which inhabited the Swell long before white men came to this desert and was one of the first to discover the ancient Pictographs and Petroglyphs of the North Temple Mountain Wash. Often he was known to sit upon the porch of vision with Andy Denney while the white-haired legend shared keen insights into the nature of his vision. Migliaccio grew!

During the vanadium rush of 1942 Lawrence Migliaccio is reported to have shipped a fortune in vanadium ore until the Government ceased that operation.[19] As with most other miners of the Plateau during these vanadium boom years, Migliaccio had obtained enormous amounts of this mineral by mining the dumps of previous operations. In his case, the rich deposits of the Chemical Products Company of Denver's effort, many years before, were there for whoever dared the intensity and fury of this land. By 1944, Migliaccio was a very well known and respected mining figure upon the Colorado Plateau.

The lull in government need for uranium did not last too long, unless you are one who lived by the every minute upon the latitudes of the Plateau. There each instant of a man's experience is an eternity. The forces of nature within this environment offer little time for slumber.

Late in 1944, the United States government found itself needing an extraordinary amount of high grade uranium ore in connection with a project to end the war. By this time the Metals Reserve Company had, in essence, turned into the Manhattan District. This new agency was responsible for taking uranium and unleashing its fury into the atmosphere. It is reported, after the energy was released over Japan, that the agency evolved into the Atomic Energy Commission.

By this time Lawrence Migliaccio had uncovered many rich deposits of uranium ore upon Temple Mountain and with the announcement of the government buying program was fast at work mining the highest grades of uranium he had uncovered. Still, the ore was all hand drilled and carried off the mountain to begin its long journey to the processing mills often several days' drive across the desert. Little remains constant upon the Plateau, therefore, a trail over the desert which existed Tuesday may have vanished by Friday afternoon. The intensity of the earth below the wheels of early ore trucks often caused the tires to melt to the rim. The sojourn from Temple Mountain, Utah to Naturita, Colorado, with a truckload of uranium ore was often the journey of life. But Migliaccio drove those miles and scaled his mountain as the war drew to its intense conclusion. The conclusion made possible by men who faced the intense force of the Colorado Plateau one stick of dynamite at a time. And the earth was not upset!

With the war's end, Migliaccio and most other miners associated with uranium ores were left to fend for themselves as the government had ceased being interested in the development of this new potential from within uranium ore. But Migliaccio continued developing his north side of Temple Mountain and was a frequent visitor to the slick rock country of Colorado at his famous Sarah Ellen mines.

BRENDA MEMORIES OF TEMPLE MOUNTAIN

Brenda Migliaccio and Sally her mule, Temple Mt. Utah, 1940's.

THE LAST MINERS OF TEMPLE MOUNTAIN, UTAH

"As my mother, Marie, loaded a bushel of tomatoes in a 1941 Dark Blue Chrysler and my Dad, Lawrence Migliaccio, filled the last of two huge water barrels with water, we were off to the mine. Dad and I in a Dodge truck with Mom following behind in the Chrysler.

It was a hot summer day and we were on the way to a hot desert called the San Rafael Swell located around one hundred miles South East from Price, Utah, always stopping at Hal Andersons Standard Station in Green River, Utah. This was always a happy time for me because I loved to go to the mine with my dad. I wasn't in school yet so there was no reason for mom and I to stay home. The extra water was for my baths and me. I just loved getting dirty and looking for rocks with the Turner kids. We played in and outside of the mine during working hours and at the cabin during morning and evening times. Temple Mountain was so beautiful to me then, and it is still beautiful to me know. Tom Watkins, the cook, was always complaining in a joshing way that I used all his water and was always teasing me about the baths I took at night in a old style wash tub. I loved riding my mule Sally and always loved to take care of her when she was not working pulling out the uranium in a custom cart built especially for that job. She was a sway back and was good at her job. I loved her allot and was always anxious to see her when I was gone for a period of time. The crew consisted of Big John Adams, John Davis, my Dad, the cook, and a couple of other miners whose names escape me. As I look back on those days all of them were exciting and full of adventure. Dad was my hero and favorite person in the whole world. He always had time for me, was very patient teaching me the exposed formations at the Historic Vanadium Kings located at Temple Mountain in Utah plus the Historic Sarah M's located in the famous Uravan Mineral Belt in Colorado. He was so proud of his operation at Temple, and the money that he was making. The uniqueness of Uranium Mining was a conversation piece in itself. Everyone had many questions about the operation everywhere we went.

Mining was a tough business then, still is. By the time you loaded the carbide lamp and struck a small flame on your mine hat, drilled the holes to place the dynamite, blast and then load the cart by hand several times in a shift as Sally pulled the cart out of the mine, dump into the shoot to a truck below it was now dinner time. The crew and I were ready to clean up and eat dinner. After dinner Dad would light a big bond fire and we would all sit around the fire and tell stories, laugh a lot and just have a great time. Lights out early to get ready for the next days hard work. Dad would have to drive all the way past Moab, Utah to the processing mill somewhere between Monticello and Moab, Utah. Sometimes, I would go with him on the truck because at that time it was an overnight trip most of the time I would stay with mom at camp with the crew, the cook, and Sally, while waiting for Dad to return. Everyone seems to know my Dad in Green River, Utah. Moab was a Uranium town and many people knew my Dad and Jack Turner also. I grew up having friends in both places and still do to this day, all related to Uranium Mining Industry. As I grew up and started school the summer months were the only time I could get to the mine. Sally was getting older and she needed to be retired. So Dad got another mule called Big Red and we retired Sally to a field in Moab, Utah. Everything was changing and Dad had many Lawsuits that he was fighting with Continental Milling and Mining Co. and Union Carbide Corporation of America to protect the ownership of his property. The Industry was like a roller coaster same as today. Only difference, Dad was in court or the Emery County, Utah Recorders Office in Castle Dale, Utah as I still do today with Recorder Dixie Swasey, most of the time. Those 1950's and 1960's court days were not very much fun because of the disappointments and heart aches in court and in the industry. Seemed like everyone wanted a piece of the action and all had a new idea of what to do to get it. Dad hired a huge Law firm in Salt Lake City to protect his interests. They got most of the money and the percentage that was paid every time we went into production until my Dad and Mom passed on.

Lawrence Migliaccio, Sally, and Camp Cook Tom Watkins, 1940's

I would inherit the Vanadium Kings at Temple Mountain and the Sarah M's in Colorado after my Mom passed on. It was up to Mom and me to make it work by a sale, lease, or mine. When the Market came back in 1976, I was ready to take on the responsibility of putting the Vanadium Kings back into production. When it was profitable to mine along came a tall skinny kid by the name of Jim Keogh from Moab, Utah after a lease. I had turned down a couple of Companies already, but the skinny kid from Moab convinced my Mom and I that they could do this. So we gave the lease on both uranium properties to a group of young guys from Moab, Utah that knew the industry, Temple Rock Mining, Jeff Jacobson was a 2nd generation Uranium Mining Son. He really new the business and how to mine Uranium. The Lease was signed and percentages given to the Energy Fuel Inc. for money distribution. The guys started hauling in all the equipment needed to start up operation. The summer of 1977 proved exhausting, since I had children of my own I was only looking at summer months to monitor the operation. I ended up living at the mine from May until school started in September. No Sally or Big Red, everything was automated that could produce many more tons during a shift. Uranium was up to $48.00 a ton, which made it economical to mine at those prices. Considering in my Dad's time it was economical to mine at $8.00 a ton.

My two boys Nicky and Michael were having the time of there life. When I was there age I had Sally the Mule, they had motorcycles, it was a paradise for them. We mountain climbed down the wash, looked for arrowheads explored the Reef, found water holes that were large enough to swim in, found Indian writings and arrowheads. They really learned the environment, the insects, pygmy rattle snakes, bats and animals involved in this adventure. Buffalo were still on the Swell when I started going with my Dad in the 40's, now, it was grazing antelope and cows. We spent some time at Denny's ranch too. That was the lifeline in the 30's and 40', but, was just a memory to me and a story to my children. The stack to the fireplace and the windmill were the only relics left at the ranch. I had plenty of stories about Mr. and Mrs. Denny to tell my kids. The old road through the Carmel Formation, was behind the ranch on the North side. The paved road now is on the south side. I remember it took at few hours to get from the Hanksville Road to the north side of Denny's ranch road to North Temple Wash. You were lucky not to get stuck in the sand several times before reaching North Temple Wash. The main road was just as bad as far as getting stuck. That first road was built with a chain and a tree dragging behind a Uranium truck, I explained to my kids. So have some respect for the paved road. At the end of the day I cooked dinner for my kids on an old stove that was left there during my dad's time. After dinner we built a huge bond fire, crew and kids sat around the fire and laughed and told stories. It was a repeat of the past all over again. As Temple Rock Mining started to produce I left my roots and bought a house in Laguna Beach, California.

I really don't know what will happen to the uranium industry at this time. I know that Temple Mountain plays a huge part in my life and it will be hard to let it go for any reason. The Mountain has a special place in all of our hearts. My Dad, me, and my sons. Over three generations and thousands more of our memories, and emotional ties to this famous historic place we call home."

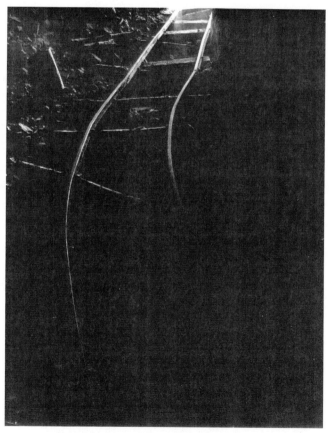

Martin Image, Tracks leading to the throbbing heart of Temple Mountain, 1976

MEMO STORY STARTING 1957

"I left Price in a hurry that afternoon, I was always in a hurry those days with my busy schedule. I was driving my White 1957 Thunderbird 400 to all the mining camps, teaching the mining camp kids how to dance and perform. Since I was a roll model, most talented in the state at that time. I held the state title of Miss Utah National Guard and attendant to Miss Utah. Everyone knew me. My car was like driving an advertisement that I was in town. All the little kids wanted to be just like me, and some mothers named there kids after me. I loved being popular and it was so much fun performing at East Carbon City, Utah for Kaiser and Geneva Steel and most events. Then up to Hiawatha, Utah for another mining convention. I performed for everyone including the Wool Growers Associations to the Night of Pithians. I raised money with shows to help build all of the Mormon churches in Price and Emery County. Danced on Television in the Original Ted Mack Amateur Hour, to local Utah's Eugene Jelesnick Talent Show Case that I won for six months on a local Salt Lake City Television Show.

Martin Image, Brenda Migliaccio Kalatges

Teaching performing arts in the mining communities was so rewarding. I even taught Michael Landon's wife, Cindy Clerico Landon. Cute little girl with a great story. I remember the day Cindy; her mother Marilyn and dad Richard came to tell me good by. They were on their way to Hollywood to enter the big time. Cindy's mother was beautiful and she wanted to try to get into pictures. I think she went into real estate and Cindy did all right for herself.

I put on many shows in Price, Utah. My first show was called Circus Time held at the Price Civic Auditorium 1958. The new Geary Theatre was located on the Carbon College Campus where I put on all my reviews until I left Price, Utah.

I was known for my professionalism in presentation and choreography. The Kids always remember their costumes, dances and pictures. I did a show every two years including my Salt Lake City studio kids would come to Price to perform. I owned a women's ready to wear store in Price so all my fashion shows had dance and up to date fashions in them. With a failed marriage, and Uranium production checks coming to my mom and I. I decided to leave the state of Utah to raise my children in California. We ended up in Laguna Beach, California a beautiful Lagoon in Southern California. I opened a Dance Studio and started to teach. I became very successful and my first show was a hit. I did dances through the years.

By 1979 I was asked if I could break a worlds record. Of course, I said. Without thinking about it. So I spent a month preparing to break the Guinness Book of Worlds Records for ABC Television, David Frost, in conjunction with David Paradine Productions. Well I did it and I felt I earned the win.

Any Worlds Record is a hard accomplishment, mine was for the largest number of tappers dancing a production number. I did it all from the logistics of the buses and where to pick up dancers. To the costumes and taught my choreography, then cut music Crazy Rhythm by Meco. Everyone bet against me, but that made me more determined to break that record and bring it home the record from Nick and Roy Castle London BBC Television London.

My reputation grew in massive production. I did a World Premiere at the Cinema Edwards Theatre at Fashion Island, Newport Beach, California, with 300 dancers dressed in pink and 3,000 poinsettia's for the movie City Heat starring Clint Eastwood and Burt Reynolds raising money for the Diabetes Association.

Not long after that huge event for the Diabetes Association I was asked to perform at the Beach for the first MTV Beach Bash. I did the promo's and performed in front of the Historic Hotel Laguna on a strip of cement. I picked my dancers and we tapped to Rocket by Herbie Hancock. It really was a hit and I enjoyed being interviewed by Mark Ubank and Virginia Madsen.

Johnny Green Years

"The most bizarre mining venture I believe I have ever been involved with afforded me the honor and privilege of meeting, knowing, and managing Johnny Green's Historic Golden Chest Mining Property in Murray, Idaho for fifteen years. I met Johnny and Bonnie Green in the early 1980's while my Vanadium Kings Uranium Mine was in production. Johnny was MGM Musical Director and Executive in charge of Music at MGM Studios during The golden years. In 1973 he was elected to the songwriters Hall Of Fame, his all time great include "Coquette", "Out of Nowhere", "I Cover the Waterfront", "I Wanna be Loved", "The Song of Raintree County", and "Body and Soul". He has been nominated 14 times and won five Oscars. His landmark original score for Raintree County earned him one of his nominations. Awarded for his musical work in movies he has four including "Easter parade", in 1951 "An American In Paris" 1953, and "West Side Story" in 1961. He was nine times musical director of the Academy Awards Show's. His last presentation was held in San Diego, California. Notes by the composer and the parallels and contradictions called "Mine Eyes Have Seen". I had attended many of his scheduled events with him and Bonnie. He was a graduate from Harvard University and one of our top composers of the 20th century.

Because of our association, when the 1986 Golden baton Cultural Art Event sponsored by the Orange County Philharmonic Society came around, I was asked if I would invite Johnny Green to attend, and perform to raise money for the Fine Arts Center in Orange County, California. I asked, and he graciously accepted. I was on the entertainment program performing a tap dance called "Let Yourself Go" from the movie Royal Wedding in which Johnny's fingers had played that boogie. I also presented one of my massive productions outside the new Hilton Hotel in Newport, California to greet and honor Mr. Green. This was a huge Black Tie event that raised untold thousands for the Orange County, California Fine Art Center, and scholarships for young deserving artists.

I continued to manage the Historic Golden Chest Gold Property until 1994 then bought the property, held ownership until 2002."

A Last Miner of Temple Mountain
Jeff Jacobson

"IT was the spring of 1977. I would be 26 that year and I'd just quit my job as underground geologist for Rio Algum. July 4th of that year some friends and I went to Telluride, Colorado to watch the fireworks and chase the girls. I ran into an old friend of mine from Moab-Jim Keogh. He had been working in Grand Junction, Colorado cleaning up the old Climax uranium mill. In one of the areas of the mill Jim said he'd found about 2000lbs of yellowcake and sold them to the Atlas mill in Moab for $60,000.00. Not bad for a days work! The party was on! We drank whiskey for three days and at the end of the third day we'd convinced ourselves that what we needed to do was look for a good Uranium mine to lease. The next day we hit the road to do just that. Our first stop was at the La Court Hotel bar in Grand Junction where we'd heard that a lot of the old uranium promoter's hung out. We went into the bar and the first thing we did was buy the house a drink and then another one and talked uranium with anybody that wanted to talk. We made quite a few new friends that day. Free whiskey will do that. One of our "new" friends told us he could put us in touch with a good mine to lease. His name was Jim Talarrico I do believe. He did just what he said he could do. He gave us a name and number of some people in Price, Utah.

Martin Image, North Temple Mountain, Utah, 1976

The mine that they had was a good one he said the claims were called the Vanadium Kings and the person that owned the claims was Marie Migliaccio and her daughter Brenda. It took us about a week to get to Price because of the fact there was still a lot of whiskey to drink and a lot of thirsty folks to drink it. We made it to Price, Utah toward the end of July of '77'. We didn't call beforehand and just showed up at Marie's door. Her daughter Brenda answered the door she had just returned from California if I remember right. We asked to speak to Marie and Brenda wanted to know what we wanted with her mom. We said that we wanted to lease her mine. Brenda liked the sound of that and asked us in and went and got her mom. When we left Price that night we had a commitment from Marie and Brenda to lease their Vanadium Kings.

A week or so later we were back in Price to sign the lease. With the lease in our pocket and a smile on our faces our next stop was Denver, Colorado to borrow money on our new lease. In Denver we went to see Bob Adams of Energy Fuels Nuclear. Bob was going to build a new Uranium mill so we figured he might lend us some money because a new mill demands a lot of feed and we wanted to feed it. Bobs office was on the top floor of the Bank of Colorado so we went right up there and asked to speak with Mr. Adams. We hadn't made an appointment and his secretary told us to wait and she would see what she could do, Bob was in a meeting at that moment. Every one in the office was dressed for business and Jim and I were dressed in old cowboy boots and I think I had on my "Keep on Truckin" t-shirt and blue jeans. Ten minutes later we were ushered into Bobs office where we met Bob and his VP, Merl Vincelett and one of Bobs head geologist. We were young but we weren't dumb and for every question they had we had an answer. At the end of our meeting Bob wrote us a $50,000.00 personal check to get us started on our mining venture amidst the objections of his VP, Merl Vincelet. Merl's objection was that we'd just spend the money on "strumpets and whiskey" and we hadn't signed anything for the money. Bobs answer to Merl was that he trusted us to do what we said we'd do. When we left we were smiling all the way to the bank which after all was on the first floor. Merl was right we did spend some of that money on "whiskey and some high priced strumpets", but by the middle of august of '77' we'd put together a decent mining operation, complete with an underground loader a couple of "new" 5 ton Young buggies a "new" loader to load trucks up top along with a wild assortment of jack hammer's, air hoses, compressor's, jack tanks, and a 30kw cat generator. It's expensive setting up a mine and before the end of August we were back at Bob Adam's door looking for more money. Bob was impressed with us and we left that day with another check but this time it was for $100,000,00. This time though Mr. Vincelett got his way and we had to sign all his paper work. We signed a good contract that day. The $100,000.00 got us 4 miners, a cook, lumber for a cabin, and enough powder, caps and fuses to blow all of us to hell and back and Jim and I still had $50,000.00 left over. Temple Mountain and the Vanadium Kings were good to us! In October of '77' we made our first shipment of ore.

T. Protopappas Image, Migliaccio Cabin, North Temple Mountain 1976.

It was a 1000 ton lot of .13% U308 and 1.00% V205. the check we got for that lot was $45,000.00 for the U308 and $60,000.00 for the V205. Hell we made more money off of the Vanadium then we did off the Uranium. Yes, we signed a good contract with Bob Adams and Merl Vincelett that day in Denver. It was good for Jim and I, not so good for Mr. Vincelett, out of the $105,000.00 from the ore lot we had to pay 20% to Marie and Brenda and Mr. Vincelett and company got $3,000.00. He had failed to proofread the contract that day in Denver he was in to big of a hurry to make us sign something. The contract we "all" signed that day called for a payback of only $3.00 a ton-$3.00 x 1000 tons=$3000.00 and Mr. Vincelett was not a happy camper when he realized "his mistake". In a 4 month time span Jim and I had went from a drunken dream of procuring a good uranium mine, to picking up the lease of the Vanadium Kings at Temple Mountain, setting up and operating the mine, selling our first lot of "ore" and we hadn't even went through a case of dynamite. You see, when we cleared out the mine so we could get to where we wanted to start mining, we found a lot of the old drifts had been stuffed with waste by the miners before us! Their "waste" was too low grade to sell when they were mining but in 1977 all that "waste" that the old timers had tried to get out of their way by stuffing it in unused drifts became a good payday for us. We did all of this in 4 months!! In 1977 we didn't have to get any permits or post any bonds or wait for any employee of the government to give us permission to mine—We just went out and did it!! Today what we did in 4 months would take a good year if you were lucky and didn't piss off one of the hordes of government employee's that you have to deal with! That's why nothing ever gets done in a timely fashion anymore! We mined the Vanadium Kings of Temple Mountain for another few years until the "Three Mile Island" scare that broke the back of the uranium market and drove all of us mine owners, prospectors, promoters, miners, mine supply company's, Mills, and all that depend on them out of business. The time I spent at Temple Mountain was not always a good time. The summer's were really hot and the gnats(no seeums) were so plentiful you could see them and they would just about drive you crazy and when it rained you were in mud up to your knee's, which made it particularly hard to do any work outside of the mine! The winters were frost bite cold, snow up to your waist and on a warm winter day there was the mud again up to your knees, which made it particularly hard to work outside of the mine! The fall was always an intensely nice time of the year at Temple Mountain. The snakes, packrats, mice and every kind of critter loved the fall at Temple Mountain. That's because that's the time of the year that they all start moving into "YOUR" house, which made it particularly hard to sleep outside of the mine. The spring time at Temple Mountain was both really nice and bad—it rains a lot in the spring-lots of flash floods, and did I mention the Mud, Which made it particularly hard to get anything done outside of the mine?? And the gnats? Which made it particularly—I think you get the picture. All this is why we spent an awful lot of time in the mine and being in the mine was conducive to work, that's why we got things done. Today, the spring of 2011 I have come full circle. I've hooked up with Brenda Migliaccio again on her Vanadium Kings at Temple Mountain and have picked up claims of my own at Temple. Brenda's claims are on the eastside of the mountain and mine are on the west side of Temple. There is a bit of a uranium boom happening once again! Brenda and I along with the other claim owners on Temple Mountain are looking at doing some deep drilling on the Temple Mountain collapse(breccia pipe). Once again I believe that "Temple Mountain" will be good to us, she's been hiding another surprise for us at around 1500ft. It might take a bit longer then it did in "1977"—(did I mention the government and bonds and permits and not ticking any government employee's off; oh and getting lucky?) but in the end I think Temple Mountain will be good to us!

Martin Image, Zane Greys Factory Butte, 1976

In the aforementioned brief the following story, involving millions of dollars and undefined human dignity, unfolds:

"...Plaintiffs obtained the mining claims as a result of a settlement of a prior lawsuit entitled LOREN HUNT, et al., vs. JESSE BITTERBAUM, et al., Case No. 1713 in the District Court of Emery County. The judgment in that case filed February 9, 1954, was based on a stipulation between the parties. The described the monuments which he had placed on the ground.

It was discovered that the first call of the Brandon survey namely a "point which is located South 40 degrees 12 minutes East, 1160.2 feet from U.S. Mineral Monument No. 246X" was erroneous and that at the point so described which is the northwest corner of Vanadium King No. 3 claim, a steel pipe stake painted white was located and such stake was actually "South 41 degrees 8 minutes East, 1259.2 feet from U.S. Mineral Monument No. 246."

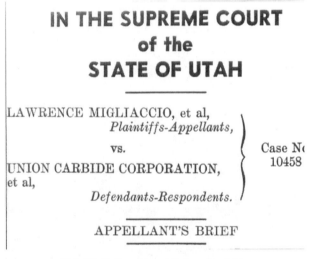

Brief/Migliaccio vs Union Carbide

At the trial and prior thereto after the discovery of the mistake, parties agreed that the steel pipe marking the northwest corner of Vanadium King No. 3 claim was not located where Brandon located it and was actually located at a point where the parties intended the claims to commence. This in effect gave all of the parties a starting point which is agreed. The north line of Vanadium King No. 3 and No. 1 was described in the Brandon survey as running from the steel pipe South 85 degrees East 3,000 feet and there were markers placed on said line and are not seriously controverted; however, the Court, in determining the north line, described it as running South 85 degrees 40 minutes East, 1445.81 feet to a car axle painted white, thence South 85 degrees 23 minutes East, 1441.7 feet or a total distance along the north line of Vanadium King No. 1 and No. 3 of 2886.22 feet.

Defendants made no claim to any property adjacent to the north line of the property of plaintiffs and do not claim anything along the East side of plaintiffs' claim, and the Court's refusal to adopt the Brandon description and order that its location be surveyed on the ground gratuitously took from plaintiffs the 93.22 feet of their claim an did not award said property to anyone.

The Court, in its description of plaintiffs' claims after refusing to adopt the Brandon description of the north line, then changed the description of the north line, then changed the description of the East end line of plaintiffs' claim. Under the Brandon description, the East end line of the plaintiffs' claim were described as follows: "South 15 degrees East 638.7 feet." The trial court described this line as South 19 degrees 8 minutes East 625.72 feet, and this narrowed the claims of plaintiffs by 12.98 feet on the call along the East end. This description of the distance from the north line to the south line becomes a crucial item of dispute between the parties since defendants own the property to the south of the property of the plaintiffs."

Remember, that our plaintiff is this man called Lawrence who was upon Utah's Sinbad with Andy Denney and the buffalo, and climbed to the top of his 6773 foot Temple Mountain to hand drill high grade uranium ore shipped by mule to the Mill at Naturita, Colorado in 1942. Our plaintiff is the man who invited Jack Turner upon the Temple for his first strike . . . Jack Turner who is responsible for showing Charles Steen where to find his first $1,000,000.00 in uranium. Migliaccio who fought the great inequities in the State's Mental Institutions and was the first ore hauler out of Temple Mountain during the forties and early fifties of the Uranium Boom. Union Carbide Corporation of America had to take a great man of your west, confine him to one small county courthouse in a valley not far from his TEMPLE in front of a practitioner of his law who took precious feet from the claims Migliaccio had pioneered 20 years before and had always shared. These feet are reputed to be worth millions and were clearly defined in Case No. 1713. The brief for Case No. 10458 continues:

"The Brandon description went, '. . . thence North 85 degrees West 3,000 feet;' the court substituted for this call 'thence 84 degrees 37 minutes West 1480.74' moving the strike of the line 53 minutes and shortening the distance 19.26 feet. The Court broke the call on the two claims into two separate descriptions, the Vanadium King No. 3 claim he described as 'thence North 85 degrees 53 minutes West 1468.7 feet to a pipe painted red or orange and in these descriptions shortened the south line of plaintiffs' claim by 51.09 feet.'

The final call on the Brandon description of the Vanadium King Nos. 1 and 3 read as follows:

"Thence North 15 degrees West 638.7 feet to a place of beginning."

The Court, in order to connect up the west line of the plaintiffs' claims, ordered a description as follows:

"Thence North 16 degrees 3 minutes West 595.5 feet to the place of beginning" not be made without resort to reference points. Some of these stakes were located behind large rocks and in areas in which Brandon could not see his point of reference."

Lowe and Cuthbert did not go with Brandon to make his survey. They relied upon him as a licensed engineer to locate the stakes in accordance with the calls as contained in the stipulation which was finally embodied in the judgment in Case No. 1713.

There was no testimony as to the location by Brandon of the various stakes which the Court now decides make the true south line of the

Vanadium King Nos. 1 and 3. The only evidence that these stakes are located in a point which was agreed to by the parties came from the lawyer, Elggren. All of the witnesses, including Elggren, testified that he did not remain at the location of Vanadium King No. 1 and No. 3 while the stakes were being placed by Brandon, but left earlier than the other parties who were there to supervise and agree upon the location of Vanadium King No. 1 and No. 3. The other witnesses testified that they did not remain at the scene of the survey by Brandon while he completed his survey. He placed the stakes along the south line and did the staking with the aid of Davis, one of the plaintiffs. Davis is unable to identify the stakes along the south line as being in the same position as they were when Brandon located them. All of the parties assumed that the staking was in accordance with the description contained in the stipulation which is the basis of judgment of the Court in Case No. 1713.

The court's findings of fact and conclusions of law and decree in the present case reduces the size of plaintiffs' claim substantially. It is agreed by all parties to Case No. 1713 that plaintiffs own two full-sized unpatented mining claims of 600 x 1500 feet . . ."

The case you have just read is unfinished today. Lawrence Migliaccio died without ever having another judge consider this matter. Do not let the Migliaccio's of your western experience wander too distant . . . alone.

POULET AND THE CURIES

One who wasted little motion upon the virgin yet energy-rich Colorado Plateau of the 19th Century, Charles Poulet (a Frenchman working closely with Pierre and Marie Curie) seemed a rather insignificant figure perched upon the red rock ledge some 500 feet above the Dolores River in the March chill of 1898. A slight man of sinewy strength, he was known by those associated with Colorado's mining industry as the one responsible for sending to Paris for the analysis of "Irish" Tom Dullan's carnotite uranium ore several years before. This essence had been brought to Poulet by Tom McKee, a person aware of Dullan's strange mineral and one who had been researching for several years to find the proper authorities to have it identified.

Balsley asserts 1st. uranium concentrating plant
in the world. San Miguel County, Colorado.

Regardless of notoriety, Poulet stood alone upon the incline whose substance vanished into the Dolores River.

Scanning a plain, some three hundred feet wide which cut through the uranium-bearing formation he stood upon, his vision of this small valley of the Dolores was unobstructed save for an occasional Pinyon Pine which, remarkably enough, drew its essence through the red rock from the river below.

The plain intrigued him for it could, rationally, be approached from but one direction yet it possessed a year-round source of water. Solitude and essence which bathes humanity, two of the three requirements essential for successful completion of his project. Obviously, he was one upon the Plateau interested in more than vernal environs that year of 1898!

Slowly tacking the red sandstone, the young Frenchman displayed the innate finesse associated with the fabled "Rocky Mountain Canary" and was soon resting on the Shinarump, amused by small brook trout facing upstream waiting for the flow of the Dolores to provide a caddis fly in distress. For Poulet's gentle presence went unnoticed by life mirrored upon the aqua plain of the Dolores River near Camp Snyder in San Miguel County, Colorado.

Descending the final short distance which separated Poulet from the Dolores, he accidentally dislodged several sandstone slabs and they slid with him into the river below. His presence was henceforth known!

Fording to midstream, he turned to face the ledge at the point he entered the river and was amazed the distance he had descended appeared more treacherous than when upon it. The thought was fleeting as Poulet first faced upstream then down without diverting the focus of his vision from a yellow vein of mineral running at eye level until it vanished into the river's plain equi-distance upstream and down. After doing this several times in succession, then ceasing, Charles Poulet experienced a visual effect which pleased him. Fording back to the formation of this effect, he determined that this would serve as "the place of beginning" for his efforts on behalf of Pierre and Marie Curie.

Within a relatively short time Poulet, along with another Frenchman, Charles Voilleque, had procured the aid of several San Miguel County frontierspersons and all were fast at work building what, to this day, is considered the very first Uranium Concentrating Plant in the world.

The Curies were continually in need of radium for their work taking place in France. Rather than incur the extensive costs of transporting the unprocessed uranium ore from Colorado's Slick Rock to Paris, this mill was devised to cut that cost to a fraction of what it had been up until 1898.

Little record remains to indicate whether the venture proved financially fruitful but this Uranium Concentrating Mill must be considered the first organized attempt in the history of the United States of America to utilize the energy within halos of uranium ore. Not predicated on destructive precepts, on the contrary, Poulet's work upon the banks of the Dolores was of the highest humanitarian motive!

Mrs. Nellie Snyder, on whose land this monument was built, alleges meeting Marie Curie in 1899 at the site. Madame Curie had quietly come to America to study the environment in which American uranium was found and to work with Poulet and Voilleque in the new facility. Her first response to the slick rock is in question. But as she traversed the deserts of Southeastern Utah on horseback and explored over the border into New Mexico, this earth could have only led to her clearer understanding of the intensity with which she was working!

It had taken the French to rive the boards from native Colorado timber to erect this first edifice used to draw "essence" from uranium ore in America. As the 20th Century dawned on the American energy experience, the next major effort at understanding and applying the "atom's" potential was being born in the minds of several within America's scientific and medical societies, not military.

Denver, Colorado of 1903 bore little resemblance to that thriving metropolis today. That is, with one exception. Since its inception the town has allowed those who chose to inhabit its boundaries two starkly contrasting visions. Gazing west from any point on Denver's plain one is humbled by the extraordinary aspect of the eastern slope of the Colorado Rockies rising to over 10,000 feet above Denver's mile high elevation. From the same spot, but turning east, a great level plain runs far into Kansas. Therefore, since the formation of the Colorado Rockies 60,000,000 years ago, Denver has been a place for persons of vision.

By this year of 1903 most documented work connected with isolating radium had been the result of European scientific and medical inquiry. Some "application" of this "inquiry" had taken place in America with the X-ray therapy on gastric cancer by Dr. H. Preston Pratt of Chicago, alone upon the banks of Lake Michigan, as early as 1896. Also, it is remembered that Nathaniel Hill's smelting operations had touched pitchblende uranium some twenty years before in Blackhawk, Colorado but few Americans had "probed" for rudimentary discoveries concerning the atom's structure and essence.

A new chapter of this history was to begin as July turned to August at the foot of the Front Range. A package had been sent to a Denver physician from Europe, the likes of which had never before been seen west of the Mississippi River. The recipient of that precious parcel was a young, unassuming, Swedish-trained American doctor.

George Stover had anxiously been awaiting this "arrival" many months. For being shipped to Denver was the first radium chloride ever to reach the West from Europe.

Immediately after the radioactive substance touched Denver, Stover began treating surface cancer with the chemical and his results signaled a need within America's medical community for sizable amounts of radium. No technologies existed in this country to produce the substance. Therefore as with the Curies, all of the uranium ore mined in America had to be shipped to France or Germany for enrichment. American industrial interest had not yet developed the technologies essential to refining natural ore.

However, it should be remembered that the only point of reference for the uranium processing industry before 1900 was still Madame Curie's mysterious mill on the banks of the Dolores River, far from human thrust, hidden from all but the most discerning eye! If the cost of shipping uranium ore was too prohibitive for her resources and she found it cheaper to have a mill built in this country in 1898, imagine the cost of shipping American uranium to Europe and back!

With Dr. Stover's early treatment of cancer by applying radium chloride, the need for a new domestic industry related to processing uranium ore was evident. However, it was to be almost ten years after Dr. Stover first began his treatments that this new industry, directed towards processing and studying the atom, was to evolve further.

As with most events within our American experience which have affected and helped direct the course of world history (for few question the fact that when all segments of American society work in unison and harmony to develop or promote an "ideal", the resulting effort can benefit existence on this planet), the formation of the National Radium Institute was a joint effort by several significant American "societies" to develop the processing technologies essential to the growth of our rare metal's industry.

High Grade Uranium Ore note Hercules Powder boxes.

By 1913, the United States Government could no longer neglect and ignore the need for radium bromide and chloride to be produced in this country. The task of developing the industry necessary to achieve this goal fell upon the United States Bureau of Mines. The heads of this agency pondered the assignment for a short time and then procured Dr. Richard B. Moore, a physical chemist of national repute, and Karl L. Kithil, a mineral technologist, to procure the highest grade uranium ore they could find on the Colorado Plateau.

In order to carry out this request, Moore and Kithil had to sojourn far into the frontier of Utah as well as the extensive and treacherous Slick Rock country of southwestern Colorado. In 1913 the environment had changed little since the Talbot Brothers had first discovered the yellowish outcrop of "strange" ore on Roc Creek during the summer of 1879. Bands of renegade Ute Indians were known to exist in dwellings far from any trails, high upon the sheer ledges of sandstone which rose thousands of feet above the Dolores River. Across the state line in Utah, Robber's Roost, to the west of Moab, was the known "hide-out" of desperados, legend today. If one did not travel on foot, the horse was the other sole mode of transport. By either mode, each mile was virtually virgin.

Throughout the spring and into early summer the two scientists, and their associates, explored the Paradox Valley and followed the course of the Dolores and San Miguel Rivers. At night they camped under the most spectacular of all heavens and, all too often, were awakened by Black Bears pillaging the campsite. They sojourned charting and mapping environments known to show traces of uranium.

By mid-summer Moore and Kithil had crossed into Southeastern Utah and observed uranium's presence on the San Rafael Swell and in the Henry Mountains to the south. The heat of the Swell proved too intense for further developmental exploration in the summer of 1913, but the historical development work by Moore and Kithil had been completed.

Turning their findings over to the Bureau of Mines offices in Denver, the United States Government was now responsible for applying the information painstakingly accumulated by Moore and Kithil during the preceding months. An answer must be found and found fast! For by 1913 the need for radium by the medical and scientific societies of America was nearing a critical stage.

The survey of Dr. Moore and Kithil had revealed that the highest concentration, then known, of high grade carnotite uranium ore was located in an area roughly fifty miles long and thirty miles wide. Moore described this area:

From a point slightly east of the junction of the Colorado River and Disappointment Creek, on the south, through a point six miles west of Naturita, thence due north to the San Miguel River, to the western boundary of the La Sal Mountains, which extend north from Uranium to Gateway.

In essence, the very soul of Colorado's Slick Rock Country!

During 1909 the Standard Chemical Company of Pittsburgh, Pennsylvania had constructed a small experimental mill to concentrate uranium ore on the Colorado Plateau. This edifice was located very close to the original claims of Irish Tom Dullan at the head of the Paradox Valley. The Standard Chemical Company was owned by two brothers from Pittsburgh, who were undertakers. Little documentation remains on these brothers, the Flannery's. Nonetheless, it is reported that because of a cancerous illness in their immediate family they sent the "Standard" west in search of radium.

With such far-reaching vision, the Flannery's are credited with eventually bringing the Bureau of Mines and America's foremost authorities on X-ray therapy together. These authorities included Dr. Howard A. Kelly, a cancer specialist from Johns Hopkins University, Dr. James Douglas of the General Memorial Hospital in New York City, and several associates. Above all others in the United States, these two doctors were aware of the need for moderately inexpensive uranium ore-processing methods in this country.

Thus, in 1913, the United States Government and American medical societies were to unite in the quest of harnessing the "atom's" eternality. The Bureau of Mines agreed it would furnish the technologies necessary to develop processing facilities if the physicians would put up the necessary capital. The physicians agreed, as long as they would be assured a requisite ore supply could be guaranteed. The National Radium Institute had been formed!

Howard A. Kelly was named president of the newly-formed Institute. Dr. C. L. Parsons was given responsibility for the investments associated with the project. Dr. Richard B. Moore, who had charted the carnotite deposits for the Bureau of Mines, headed the development of all mining and ore concentrations along with Dr S. C. Lind, a pioneer of man's search for the "essence" in uranium. Moore's associate while charting the Slick Rock and Swell, Karl Kithil, was chief mineralogist. A mining engineer by the name of John A. Davis was named superintendent. Surely six men of destiny!

The Institute came into full being early in October of 1913 and Davis immediately set out to erect two concentrating mills. The first appeared in a small, red canyon some fifty miles from the nearest center of commerce, Placerville, Colorado and sixteen miles from Naturita. It was this spot that old-timers in the area named the "doctor's camp." With the aid of frontiersmen from Western Montrose County, Davis faced the elements and won. The mill was refining the yellow carnotite procured from outcrops nearby before the turn to 1914. The second mill was erected in South Denver and was in full production by June of 1914.

Do note, however, that during this same time a company from Denver, the Chemical Products Company, had been fast at work, to the west of the San Rafael Reef and had isolated radium from Temple Mountain's ore.

The National Radium Institute operated seven days a week and was never known to cease. During 1917 Dr. Kelly and the General Memorial Hospital found they had procured enough radium for the foreseeable future and the Institute was closed, the corporation disbanded; but not before the potential of the "atom's" energy had been probed by an arm of the United States Government.

With the Government divorced from the development of atomic industry in 1918, it was to be many years before a vision of truly "united" effort would unfold in regards to the exploration and exploitation of uranium's energy. Yet to this day, little in life is more essential than "united efforts" at stopping cancerous growth. Therefore, little has changed in this regard since 1917.

If one must, seek the quietude of the San Rafael Swell or ascend high into the virgin meadows of Eastern Montrose County, Colorado and listen! Place your ear to the earth and remain receptive! To those who will be still enough a lyric of Aeolian structure shall be carried on the wind. A plaintive lyric calling mankind to once again unite in study of the atom's essence. For cancerous growth does exist. And is not cancer but the atom run awry?

Howard Wilson Balsley, Moki/Anasazi Ruins, Grand Canyon, 1911

YOUNG FOREST RANGER

The winter months of 1917 did signal the direct loss of interest by the United States Government in the development of atomic potential. But those of vision within America's energy experience would not be driven from the isolated extremes of the Four Corners. For even with the disbanding of the National Radium Institute, many who had experienced the Aeolian lyric of Colorado upland meadows were pensively probing Mother Earth.

Summer's heat gently quilted the southern slope of the Blue Mountains some 16 miles west of Blanding, Utah in 1917. Its presence pressed down upon a young forest ranger from Moab even before the morning sun had illumined the east.

Day and night often run together for the humanity to set upon the Great Southern Utah Desert, as each instant is often a cosmos and Howard Balsley had not rested very well the several hours he had been still after darkness descended upon his camp on the "Blue's."

Balsley had been sent by the United States Government to classify as much land with agricultural possibility as he could discern within the boundaries of the La Sal National Forest Assisted by a soil expert from Washington, D.C., he had spent the summer probing the earth with an auger and collecting a thousand samples of Colorado and Utah strata. The previous day he had been informed a group of renegade Utes had "quit" the reservation and settled into Allen Canyon. Moreover, the Utes were trying to raise corn without water. Balsley had been asked to work with the band and develop an adequate irrigation system, as Allen Creek was known to carry water year-round but a short distance from the area being sown.

Kicking the sage embers, remains of last evening's fire, he placed several "cow chips" upon the coals and a strange, pungent odor soon permeated the small campsite. The scent of the fire seemed timed with the first signs of morning light as Balsley compagno was aroused.

Fully aware that the Utes in Allen Canyon were several hours away, Howard Balsley felt it imperative to be on the trail with muted light! For here was a man eager to make contact with those who were first upon the Four Corners, and he knew all too well that the lessons he would be taught should far outweigh the effort of his assignment. Each such contact he had made while surveying the La Sal had deepened his respect for the earth and those who have learned to live harmoniously upon it.

Picking their way through the sage and loose shale shielding the south slope of the Abajo Mountains, the rangers marveled at the stillness which surrounded them. On a small crested butte Balsley rested and, facing north, was drawn to a cloud of red dust approaching the formation. The haze appeared to be several miles away, yet each minute that passed issued it closer to the rangers. Some short time later the vision of an Indian could be clearly discerned, leaving both rangers a tad taken back by the intensity of the figure which continued forward. Fifty-nine years later, Howard Wilson Balsley continues the story:

An Indian rode up, pulled his horse to a stop, leveled a cocked 30-30 . . . I was on the opposite side of my horse and, by golly, I had often wondered what I would do if I got badly scared and I sure found out, but in spite of it I held my ground. There just wasn't anything else to do.

The man who was with me had ducked behind a rock and the Indian never saw him but he was sure swearing at me and every once in a while would cock the gun and aim it at me. I tried to tell him that we weren't trying to take their land but was trying to help them irrigate the land that was planted . . .

A slight chill from the canyon below had reached Balsley and, upon reaching him, chilled him. The Ute remained for what seemed an eternity and rarely pointed the 30-30 away from the lone figure of the ranger. Suddenly, in as much fury as he had arrived, the Indian turned his horse and vanished without issuing a shot. It was later reported to Balsley that the mounted Indian had already killed four persons before encountering the rangers near Allen Canyon . . . Just one day in the summer of 1917 for a seeker of uranium's essence.

Twenty-five autumns were to act as a buffer, separating intense earths from intense heavens, before the United States Government would once again act in true union with American industry utilizing and studying the potential hidden deep within mystical uranium ore. While those autumns blanketed the Swell and Slick Rock, the only true joint effort at understanding this "potential" within the American Experience appears to be the union of visionary individuals and their earth. For even with the disbanding of the National Radium Institute there were those to continue caressing the Four Corners one rock at a time!

One of uranium's byproducts, vanadium, was the property which kept this energy industry, so essential to mankind's history, alive for the next twenty-five years.

Let no one surmise the years of vanadium's development were the least bit "easy" and lacking in vision and feats of success though pitted against tremendous environmental obstacles. The San Rafael Swell and Colorado Plateau had not ceased being extreme and that intensity still pressed down upon each individual who traversed and lived upon those lands! The winter of 1932 was the most passionate in history. That fury is legend to those who fronted winter's wrath at the United States Vanadium Corporation's operation in Rifle, Colorado.

American governmental thrust was totally directed towards bringing the "Depression" to an end and working with agencies and industries a little more essential to the American "way of life" in the 1930s. Little effort was devoted to promoting and developing uranium's energy producing potential even though the public of our great land often traveled the Galaxies with Buck Rogers and Flash Gordon (powered by nuclear fuels) each Saturday afternoon while the reality of the l930s became history.

Two prospectors San Miguel County, Colorado, 1918

In fact had not a "union" of corporate thought kept vanadium's aspect of uranium's essence alive for many years, the intensity always associated with atomic potential may have proved too overwhelming for even the most well-equipped individual. Yet who can judge, for as corporate union developed nuclear potential, independent and individual experience was often a step ahead. Regardless of who was doing what upon which earths within the realm of the great Plateau, all were involved in harmonious development.

Howard Wilson Balsley is legend for his vision and efforts towards developing uranium's potential. He is the history of so much of this development that one can not contemplate very many instances within the evolution of our uranium industry since 1907 without "thinking on Balsley." He stood as a pillar of independence upon the Colorado Plateau during the years when agencies of the United States Government were not to be found supporting American industry in promoting the atom's eternality. But Balsley was not always alone! By 1931 the Four Corners still attracted others of independent nature.

While corporate thrust gleaned what success could be from the vanadium ores procured and processed in the western United States during the l930s, independent persons of the earth had not left the canyons and buttes of Southeastern Utah and Southwestern Colorado. It should be noted that during the l930s the producers of uranium ore had practically no market for the uranium content of said ore, except for Howard Wilson Balsley, who had warehouses in many locations throughout the Four Corners. In fact, with virtually no market he still maintained warehouses and other storage arrangements in Blanding, Monticello, Moab, Cisco, Thompson, and Green River, in Utah and Grand Junction, Newcastle, Meeker, Montrose, Naturita, Dove Creek, and Egnar, in Colorado. So, for the humanity which could not escape the lure of uranium's potential, Mr. Balsley afforded some shelter for many solitary promoters of the domestic utilization of atomic properties. But who would have considered setting upon the isolated reaches of San Juan County, Utah in 1930, to keep uranium's pulse alive?

The frozen fury attributed to the winter of late 1930 and early 1931 (when the upland meadows of the Western dope of Colorado's Rockies rest blanketed with blizzard and dry snow) had little effect upon Benito Sanchez and his flock of sheep which were foraging on the fine green grasses of Cottonwood Creek some fourteen miles west of Blanding, Utah. Several months before he had driven his flock from those reaches, now frozen, to the warmer climes of Southeastern Utah. That land of apparent death which, remarkably enough, gives life to persons who will humble themselves to the reality of this paradox. The humble Mexican sheepherder was such!

While relocating his camp in March of 1931 Benito discovered an outcrop of extremely yellow ore near a spring of precious fresh water. Thinking the substance may be gold, Sanchez loaded his saddlebag with the precious stuff and carried it with him the next time he traveled to Blanding. While in Blanding he took the samples to his employer, Thomas A. Jones, a pioneer personality in Utah's San Juan County history. With obvious excitement Jones then took the ore to Arah

B. Shumway. It took little time for Shumway to report the ore was "carnotite." Several years before, Shumway had mined uranium in Dry Creek Valley, a valley whose uranium potential was first marked by Dr. Moore and Karl Kithil in 1913 while mapping for the National Radium Institute.

Neither Jones nor Sanchez showed any interest in mining the area if the ore was not gold. In fact, Benito offered to tell Arah Shumway where the deposit was located and was only interested in a small interest of any claims staked there This suited Shumway just fine for he was curious about deposits of carnotite being located so far away from then existing activity. Shortly thereafter Arah Shumway and his brother Harris retraced the sheepherder's trail back to Cottonwood Creek, but were unable to locate any signs of carnotite. However, returning to the area several weeks later they prospected the canyon above the creek and located several outcrops of carnotite up and down archaic Cottonwood Canyon, Utah.

From sheepherder, to miner, to Balsley. Kindred souls, who relied on their earth for existence, were responsible for keeping America's uranium industry alive during the 1930s. And the earth did not object!

With the fourth decade of this American century about to dawn, was atomic potential solely in the vision and reach of frontierspersons who inhabited the most desolate environment within America's experience? Of course not! Had not Madame Curie headed scientific inquiry into uraniums essence before the turn to the 20th Century? In fact this grand lady was alleged known to have sojourned at the Four Corners before that turn, a sojourn in quest of uranium's warm glow.

Though often from corners of the earth far removed from America's West, mental frontierspersons had been probing for uranium's essence. Einstein, Fermi, Rutherford are more than mere echo. Such as these were exposing themselves to "the fury of the Sinbad" in vastly different climes! At the same time, small shipments of ore were traversing this earth, destination ... "Mr. Balsley."

There was little concern expressed by the American governmental community when Madame Marie Curie "passed on" in 1934. Even less concern was expressed when her daughter, Irene Curie Joliot, and Irene's husband, Frederic, discovered how to induce artificial radio activity that same year. Almost unnoticed by said community was the fact that Enrico Fermi split the first atom that self-same year. A fact even unbeknownst to Fermi at the time! Yet 1934 remains as a most significant year in the history of man's search for the energy within the atom, the atom within uranium ore, uranium ore which is generating over 20 percent of America's electricity, presently!

While select men of "thought" throughout the world began to study Dr. Fermi's effort and applied his method to their work, it had become apparent to many within scientific societies that energy from the atom had been released and could be released at the will of man! Many aware of this discovery prayed, while just as many anxiously probed deeper.

By 1939, with miners of the Colorado Plateau persistent in their quest for traces of yellow carnotite uranium ore, one rock at a time, members of the more intense, yet subtle, isolated reaches of the world's scientific community had come to the obvious conclusion: the essence of uranium could be so induced to create an energy force never before imagined in mankind's experience. There was also agreement that if human beings desired to unleash this force for destruction purposes, any nation possessing the device could dominate the world's communities. And the earth was not quite as pleased!

There was scientifically whispered to be much unrest in Europe as the fourth decade of this century dawned. Scientists had united in effort to keep Fermi's findings and the fruits of those who followed his example from the Axis powers. It was also agreed that every effort was to be made to warn the governments of the Western world, especially the United States, that this potential was now fact, no longer the musings of men possessing vision and men lured to the soul of this earth by an essence they could not explain. Scientists had finally proclaimed that uranium's essence was, in reality, the essence of all life then known to man. Just a little more intense and dense than other elements.

Thrust into the conflict of the Second World War, the United States Government continued to do little to study the theory of nuclear fission set forth by the scientific community.

However, late in 1939, the government did appoint a committee to explore the chances of nuclear fission being used for military purposes, but this committee accomplished little. Though it must be noted this was the first effort on behalf of the United States Government to explore the potentials of the atom since the National Radium Institute was disbanded in 1917! However, now the motive had changed somewhat. The concern was much more than cancerous growth!

With the United States deeply entrenched in the Second World War, by 1942 this country could no longer ignore either the "essence" or "industry" associated with uranium ore. The time in American history had come for the second "united" effort aimed at exploring and exploiting uranium's potential.

Albert Rogers, first pioneer Uranium prospector of Moab, Utah area, about 1910, Brumley Ridge, near Moab, the Springfield, later Blue Goose Claim. Photo by Lilla Winbourn.

What task is impossible to the American Experience when true union of our national consciousness and deed is applied to any problem? That is, what has not been accomplished by our nation when all work towards a common goal . . . led by the United States Government? History dictates little! The chain of events in 1942 which heralded a new dawn in mankind's search for the essence in uranium?

Driven off Temple Mountain, Utah by the snows of 1941; snows which one watches approach the highly eroded terrain of the "Sinbad Country" from the Northwest and Canada, cold, angry, wet, and eventually radioactive snows which one finds difficult to ever completely "dry off"; snows which gather their intensity from traveling thousands of miles; Lawrence Migliaccio had retreated to Price, some 100 miles north of his mountain. Price has always been the "Coal Capital" a great mining center within western American history.

This "Space-Age" Pioneer spent the ensuing winter months carefully studying his charts and maps of Utah's San Rafael Swell and assembling his resources for Spring's assault. He had uncovered several rich deposits of uranium ore the preceding year and had hand-picked and drilled over 10 tons of "high grade" from very near the summit of Temple, some 7,000 feet above archaic earths. The mere feat was already legend at the United States Vanadium Corporation's mill at Naturita, Colorado, where he delivered the precious mineral. His total attention focused throughout that winter on how to mine the mineral more efficiently.

With America at war, here was one aware of work directed towards the development of nuclear fission who felt it would only be a short while until the government would need all of the rich Temple Mountain ore he could procure. He planned to be ready! For above all else, frontierspersons of the Four Corners are always known to be prepared.

With spring's first blossom, tiny virgin buds upon the cactus, a time when the Colorado Plateau is bathed in "electric color," too much life springing forth from winter's slumber, too quickly. The husky Italian pioneer headed south to Green River, Utah, the first stop in his "assault on the Sinbad."

In Green River, a tiny hamlet of life upon the fringes of the San Rafael Desert at the foot of the coal-rich Book Cliff Mountains, Migliaccio gassed his battered red Dodge truck at "Hal's Standard." This, the last available gas for a hundred miles to the Southwest across the desert, and, where a trail did exist it was largely due to someone passing before, dragging a log behind their truck or wagon in a necessary yet often vain attempt to subdue the shifting flower-like sands of the San Rafael Desert. The desert whose sands travel north on warm winds from Mexico which glide over a place of origin, the Goblin Valley, just to the south of Temple Mountain.

As dawn's first radiance illumined the east, Hal told the pioneer that he, Migliaccio, was the first to set upon the desert this spring with the exception of Andy Denney, his wife, and several grazers. Appearing pleased with the news, he was soon upon the desert, but not before attaching a five-foot, twenty-inch in diameter, rough-hewn Aspen log to the bumper of his Dodge with the help of ten feet of heavy chain! His search for raw domestic energy had continued, strictly controlled by certain structural features of Utah's San Rafael Swell.

Martin Image, Vanadium King No. 7 1940's ore shoot.

1940'S GLIMPSE

This uranium producer proved visionary that self-same year. Naturally, with the United States at war and American industry producing at a pace unmatched in the annals of World History, uranium's essence and by-product, vanadium (which it must be remembered was essential to Allied effort of the First World War), would be needed for increasing the tensile strength of steel produced in this country!

While frontierspersons traversed the desert with log in tow, the United States Government came to the realization that governmental industry would use all of the vanadium this nation could produce, and immediately! For the first time since 1917 a direct act of the government was responsible for increased activity within the industry so related to man's search for domestic energy from within the halo. That need solely for defense of your freedom! And the earth did not object!

During the spring of 1942 the federal government created a company known as the Metals Reserve Company. The function of the agency was to control all production and processing of vanadium ore in the United States, and whose responsibility it was to see that America's increasing need for the mineral was fully met. For a short period of time those miners who had associated with nuclear earths were rewarded with an open market for their effort. Migliaccio found enough vanadium exposed on the Sinbad to allow him the opportunity to make a "fortune" by merely collecting the vanadium exposed on the north side of Temple Mountain left by the Chemical Products Company in 1917.

The formation of the Metals Reserve Company issued new life into uranium development as persons associated with the mining industry of the Four Corners were also given the opportunity to work on a steady basis, and the administration associated with any full-fledged "industry" afforded the first gleam of hope that the future of nuclear fission would be a joint effort by Americans all, concerned with developing atomic potential.

By 1942 a new generation of visionaries had been born and almost overnight the vanadium industry grew to a sudden maturity. Whereas five million tons of vanadium had been produced on the Colorado Plateau in 1941 by 1942 space-age pioneers fronted the Plateau and had extracted ten million tons.

The Metal Reserves Company proved only one of several events associated with 1942 responsible for forcing new attention to the importance of America's uranium industry.

Given the opportunity to retaliate against the attack inflicted upon American industrial resource Imperial Japanese Forces, by September of l942, the United States Government threw its full support behind those determined to control the energy from within the atom. Reports from Europe indicated that German scientists were developing a device of destruction using nuclear fission. For the first time within the history of man there existed documented evidence mankind was developing the energy from within the halo with intention not founded on the harmonious precepts of history afore recorded.

This startling slap of reality awoke the proper administrations within this nation's governmental community and the "Manhattan Engineering District" was formed to develop America's nuclear industry at a pace much more rapid than the effort occurring within the cloistered walls of the Kaiser Wilhelm Institute in Berlin. General Leslie R. Groves was appointed acting head of the Manhattan Engineering District while Colonel K. D. Nichols headed the research bureau.

Facing an enemy composed of human beings, human beings known to be so bent on the destruction of the American Ideal that each had sworn to devote their every instance to the extermination of our race, even if it took Kamikaze effort!

With the District's creation, a signal was sent throughout America's uranium industry heralding a market for high grade uranium ore. The frontierspersons who had once only been paid for the vanadium content of their ore found the processing mills upon the Plateau buying with reckless abandon. Rarely since the discovery of pitchblende by Dr. Richard Pearce at the Wood Mine in the Central City Mining District of Colorado in 1871, had those procuring uranium ore been paid for both the uranium and vanadium content of the precious mineral! An industry grew, a nation was growing, and keen minds whose sole vision was halting the death-strewn path of Imperial Force, expanded.

On December 2, 1942, as chill winds from Lake Michigan drove most inward to avoid that fury, a squash court under the football bleachers at the University of Chicago served as the foundation for the final event of 1942 essential to man's search for properties associated with uranium. Enrico Fermi, the fountainhead of research on nuclear fission, induced the very first sustained nuclear chain reaction. As visionaries had been dragging one-hundred-and-fifty-pound logs behind trucks to assure a trail for those who wished to follow into the Goblin and Paradox Valleys, Fermi had piled bricks of graphite in regular layers and placed blocks of uranium metal several inches in size within that pile of graphite.

Far from the "deserts of intensity," Donald James Hughes of the Brookhaven National Laboratory reported, in his epic treatise on the harmonious uses of nuclear energy, just what did occur within Fermi's mass of graphite and uranium:

The idea behind this rather unusual combination of graphite blocks and uranium was extremely important, and basic to the successful attainment of atomic power. The nuclei of carbon atoms have the ability to reduce the velocity of neutrons that hit them but with very little probability of absorption of the neutrons. The low absorption is directly related to the structure of the carbon nucleus, which, consisting of three pairs of neutrons and three pairs of protons, is a very stable configuration. Thus the fast neutrons emitted from a uranium lump following fission are slowed down by the carbon nuclei, without being absorbed, until they're of the same energy as the carbon atoms themselves, a few hundredths of an electron volt.

Because of their specific properties the graphite atoms fulfill the dual purpose of hindering the escape of neutrons from the structure and, by slowing them down, of making their absorption true for almost all materials, that neutrons are absorbed more and more readily as their velocity is decreased. A neutron in such a combination of graphite lives much longer than in a pure block of uranium alone . . .

While the graphite blocks and uranium lumps of this first chain-reacting pile were being put in place, the neutron intensity rose gradually until the chain reaction began when criticality was passed . . .

This occurrence on the squash court at the University of Chicago on December 2, 1942 was final proof that an atomic bomb was possible.

Armed with this equation, the Manhattan Engineering District set about to further develop the force.

The twelve months of 1942 opened new vistas in industrial evolution which has, subsequently, played an essential role in all human destiny associated with planet Earth, even though many summers have passed since "the first Aspen log of the spring of 1942 was drawn across the desert."

Two events of 1944 were to drastically affect the uranium industry in America.

The national unity of 1942 (all Americans working together to halt the aggression of Imperial Force) paid off, and by early 1944 American Industry had stepped up production of war goods to the point that demand for vanadium ore diminished. On February 28, 1944, the Metals Reserve halted all purchasing of vanadium and miners of the Colorado Plateau found themselves with little market for their effort. The Sage of Moab's Red Rock recalls the date vividly:

Well, all was going well and many old mine dumps that ran very low in uranium but fairly well in vanadium were scooped up and sold to the Metals Reserve. However, late in February, 1944, the Government suddenly discovered that it had vanadium "running out of its ears, as it were, and orders were issued to cease buying vanadium promptly at midnight on February 28 of that year. So, we were out of business again, except for my contract with Vitro. At the time the close-down order came from Washington, I had a considerable tonnage of ore accumulated over at Dove Creek, Colorado, that was too low in uranium to ship to Vitro but it had a good vanadium content. By working day and night with two good helpers, we got the last load of ore delivered to the Dove Creek stock pile at exactly five minutes to midnight, February 28, 1944.

Few choices were open to those men of the Colorado Plateau. They must turn to other industry related to the war effort or pause for a short while and calculate what the government was going to do next with this industry essential to all phases of the war effort. What was that next step to be, vanadium or uranium? Many pondered 1944 away "listening for the reply."

Martin Image, Aspect of Zane Grey country, gazing south from the foot of the San Rafael Reef at Temple Mountain, Utah. Petrified Jurassic sand dunes.

With the government's need for vanadium negligible, Andrew J. Denney was fully aware the atomic cycle dictated the next need would be uranium, once again, before the war ended. All through the summer of 1944 and into the fall he prospected the north side of Temple Mountain seeking traces of uranium's intensity. Though no great market then existed for this work he was aware his government could not ignore the potentials of the mineral he was seeking.

A visionary compared to most, Denney had long before been aware of the work associated with nuclear fission and steadfastly developed towards the day the federal government would finally fathom the realm of this mineral, uranium. While prospecting high in the San Rafael Reef which rises over 2,000 feet above the San Rafael Desert, he came upon a discovery that moved him as much as any of the uranium discoveries uncovered on the Colorado Plateau. While ascending a small fracture, to the south of North Temple Mountain Wash, he came upon an expression which totally stunned him. In the half light of an early August night in 1944 he came face to face with over eighty feet of red anthropomorphic figures painted on the face of the reef. Prospecting for nuclear ores, as America was experiencing a conflict unmatched in human history, Denney had come across pictographs thousands of years old, over fifty miles from domestic water! The figures stood to eight feet and ran for over 80 feet. The vision nearly knocked him from his feet! Within several rods the past and future of all mankind suddenly passed in hypnologic fashion. The desert night's press upon his person left little for this man of destiny to do than stare at the up-turned palms of his calloused hands. One space-age pioneer's summer of 1944 had not passed in vain!

But that eternal force, symbolized in the pictographs of North Temple Mountain Wash, was still being explored and studied by men of scientific vision as the fifth year of conflict with Imperial Force dawned.

The newly created Manhattan Engineering District was not mere lip service by 1945. In fact, this agency whose purpose it was to further study and development of nuclear potential had created three entire cities based on atomic research and exploration in an attempt to check the rapid progress of Axis atomic development.

The lush hills and forest surrounding Oak Ridge, Tennessee was inspiration for many persons sent there to study "fission." Crispness of Washington State, stimulated hundreds, then thousands, directed towards the newly created environment of Hanford, Washington, to study and develop "essence." Finally, just to the south of the Colorado Plateau, men of great vision worked on ending world conflict. The location of Los Alamos, New Mexico, legend when thought is the concern, served as the third site for the humanity employed in a project which was, in reality, born with the detonation of the atomic bombs over Japan.

As research at Los Alamos and Hanford was proving successful, the United States Government came to the realization that there existed a further need for uranium, not vanadium ore. The government was suddenly buying and commandeering all of the uranium in the nation. The final pieces of the puzzle surrounding atomic power's reality were being placed and pieced together, and the key was simply the right load of processed uranium ore. For each atom since "the creation" has been a tad different.

The Atomic Bombs were dropped and Imperial Force halted with uranium from the Colorado Plateau.

Naturally, with the ending of the Second World War, most who were close to the industry of atomic potential surmised a future of unlimited horizon as man would certainly harness atomic force and develop that potential into harmony.

Scant shipments of ore were being bought on the Plateau as Congress passed an act on July 31, 1946 which provided for the establishment of a civilian organization to replace the Metals Reserve Company. This "organization" was called the Atomic Energy Commission and the duties of the newly created organization included further development of atomic potential. Do remember, at the time American Servicemen were returning from foreign soils, some memberless, some in flag-draped caskets, some blind, many whole . . . there was little question that the force used to stop "force" was essential and necessary. As young American males were bathed in eternal, slumber washed by warm seas of the South Pacific, men of vision were losing eternity upon the force which put those thousands of young Americans to rest.

The months following the passage of the McMahan Act were used to select the five persons who would head the commission and by January 1, 1947 this task had been completed. Over forty-nine years had passed since Charles Poulet and Charles Voilleque erected the first uranium concentration plant in the world on the Dolores River at Camp Snyder where Highway 80 now crosses the Dolores San Miguel County, Colorado. Elected as Chairman of the Atomic Energy Commission to further the development of potentials first concentrated by Poulet before the turn of this century was David E. Lilienthal. Lilienthal, an epic man of this earth, had been director of the Tennessee Valley Authority, an organization also founded on man's search and study for "essence" and energy. It was only natural that a man associated with the "forces of nature" would head this newly appointed commission to develop atomic force, a mere force of nature.

The direction of Atomic Potential was not handed over to "madmen" and persons bent on destroying this cosmos. A man whose life's work had been the study of domestic energy was to head the development of nuclear force. The Federal Government had not erred in its choice of leadership. Lilienthal's assistants included Sumner T. Pike, William W. Waymack, Lewis L. Strauss and Robert F. Bacher. Nuclear potential was to live!

As the Atomic Energy Commission was finally administratively evolving, miners of the Great Plateau were finding development of the industry associated with procuring the very essence of nuclear power, the uranium industry, was last on the priority list of the commission.

The new "Energy Commission" was simply taking time to associate itself with the development and realities of uranium's essence. Yet the men of America's great West, those who had faced the tremendous intensity of the San Rafael Swell and Colorado Plateau, these persons who knew more "on" this property than even those who created the atomic reality of the Second World War, were not to let newly appointed Commissioners slumber too long!

One night upon the Great Plateau or Great Domal Uplift of the San Rafael Swell, in search of atomic force, reading geological bulletins by the light of a Pinyon fire as heat from the flaming bush touches the dry desert night, small climatic eruptions occur which issue forth a sound as resounding as the crisp response of a 32 caliber Winchester pistol shot into that same night. Those of America's uranium industry have certainly been men of ceaseless vision!

A nuclear scholar discusses one reason why the newly created Atomic Energy Commission was slow to develop an extensive program for domestic uranium production, a situation that Howard Wilson

Balsley of Moab, Utah and Fendoll A. Sitton of Dove Creek, Colorado were to soon act in an effort to rectify. Dr. Gary Shumway reports:

Martin Image, Path to Migliaccio Prospect, 1976

One of the many problems to beset the Atomic Energy Commission during the first few years was that of deciding whether to establish an extensive program for the domestic production of uranium. After examining World War II production records and obtaining the option of various officials concerned with uranium production, the A.E.C. in 1947 estimated that the entire uranium reserve of the United States was less than one million tons of ore, containing not more than 2,500 tons of uranium oxide. (Shumway's source is Harold B. Meyers.) This mistake on the part of the A.E.C. affected much of its domestic purchasing policy during the period 1948-51. The A.E.C. failed to realize that the production of uranium during World War II was not a good indication of the potential output of the area. As the miners were only paid for the vanadium content of their ore, mines that contained ore running high in uranium but low in vanadium were not worked. Moreover, many of the easily accessible high grade trees and other deposits that were rich in uranium had been mined long before the war. Thus, to anyone who was unacquainted with actual conditions, it would appear that there were only limited amounts of uranium in the region. Also, there had been little production from the Shinarump Formation during the last forty years.

Concern that this newly appointed commission, may be misinformed on some facets which would directly influence the growth of this energy potential, Howard Wilson Balsley was soon in communication with Senators Edwin C. Johnson (later Governor of Colorado) and Eugene D. Millikin, both of whom were members of a Congressional Committee on Atomic Energy in 1947. The men from the sage would be heard!

Finally, it was arranged through Edwin Johnson and Eugene Millikin for Mr. Balsley and Fendoll A. Sitton to traverse the land east of the Rockies and meet with the Atomic Energy Commission, in Washington, D.C. at noon on February 14, 1947. This meeting proved to be the first of several such encounters occurring that year. Meetings in which visionary men of America's West and East dialogued late into the night on how America was to keep the pulse of atomic potential alive within the American Experience.

As a result of that trip, Howard Balsley, Wendell A. Sitton, and Ray A. Bennett were to open the eyes of the newly evolved commission and force attention to the eternal resource within those boundaries of the land called The Four Corners. East and West had, once again, joined in development of atomic energies.

This chronicle is not composed of persons who many associated with the study of documented energy histories within our American Experience would or have considered "base." The humanity issuing forth has not included persons of perverse nature or men who desired the "energy" they were seeking for unharmonious reaction. Such do not seem to be included in careful study of the administrations within the formal structures developed by the United States Government for the evolving of nuclear potential. In fact, the opposite appears too true. Most associated with the quest of "uranium energy" have been almost metaphysical and mystical in countenance, deeply religious souls dedicated to serving mankind in some small fashion; though that fashion took them far beyond the vision of many. Those years many wandered so alone upon the Great Domal Uplift, and frontierspersons traversed the virgin forests of the Great La Sals, were not predicted on disrupting the forces of nature.

David E Lilienthal, pioneer of enormous conscience, had predicated his existence on learning the pulse of the great Southeastern waterways of America long before he was chosen to touch the pulse of the atom. The scientists or "camp of doctors" who manned the National Radium Institute's settlement near Naturita, Colorado in 1917, were pioneers in the eternal war against cancer. Howard E. Kelley who headed the "Institute" was not a man of violence as northern heavens pressed down upon his camp. So many winters working for mankind from a tent on the Colorado Plateau!

The extensive development and exploration program begun in April 1948, by the Atomic Energy Commission, opened the isolated extremes of Colorado's Plateau and Utah's San Rafael Swell for a press of civilization upon the land in search of mineral wealth paralleled by few mining events in the history of man. The remarkable uranium boom of the late 1940s and early 1950s is legend today. Yet it was brought about by the Atomic Energy Commission as a reward for the industry of "future essence" which had struggled so to maintain dignity and literally keep the pulse of "atomic force" alive within Western history.

By the end of February 1948 uranium producers on the Colorado Plateau began to hear talk of new programs being initiated by the Atomic Energy Commission, and included would be new purchasing legislation aimed at offering the miner a greater market and price. This was the first time in this industry's history (the industry of nuclear growth) that uranium producers had been fore warned that an "agency" with capital was considering such.

Anticipation ran high among those miners of the Paradox Valley and men such as Migliaccio and Ekker over on the San Rafael Swell. These persons who had been blindly groping at the earth and traveling great distances to deliver their bounty to a market which had complete control over the price they were to receive, were finally being "guaranteed a market" and good price to produce essence! For such as David E. Lilienthal are "humanitarian" business persons, who often place their earth need before mankind's (which our chronicle has implied must be done on great occasions for Americans to continue their existence). The earth is a great teacher of man! And who is more aware of this essential truth than those who have mined "halos of uranium ore" in total darkness deep within the throbbing heart of Temple Mountain? That essence a little to the North of the Goblin Valley.

When the price was finally announced by the Atomic Energy Commission, most uranium producers were dissatisfied with the figure quoted and forced the Commission to add a bonus. The Commission agreed and historical development work commenced in the uranium producing regions of America.

ZANE GREY LINGERS

"I own the most beautiful mountain in the world," kept reverberating through my consciousness as our self-contained four-wheel drive desert van slowly inched its way upon a thin path of colored nuclear earths paving the bedrock of North Temple Mountain Wash, a significant chasm in the two-thousand-foot high, one-hundred-mile long San Rafael Reef, a sawtooth ridge rising above several of Utah's southern deserts. Certainly some of the most environmentally intense and domestic energy-rich land on earth!

Al Szabo Image, Brenda at Vanadium King No. 6 Mine

Suddenly, the amber morning sky appeared to vanish into the earth as reef and sky became one directly above our dusty unit dwarfed by these mammoth Jurassic monoliths.

Navajo Sandstone walls were no more than eight feet apart as we stopped. My vibrant guide and I disembarked.

One hundred feet north we approached a shaded fissure five feet wide surrounded by lichen-covered boulders, piñon, sage and quakin' aspens appearing to ascend into a microscopic blackness. We disappeared into the cavity and began a sojourn through this petrified reef.

Roughly two hundred feet of climbing had been completed when to my surprise the fissure opened onto an eroded ledge 50 feet in diameter. Scrambling up friable boulders, my guide could look east and spy the flat Morrison top of Robber's Roost, Utah some thirty miles distant. The stillness of this morning seemed to quiver.

"It was Robber's Roost where Butch Cassidy and the Sundance Kid, along with many other desperados—including the Wild Bunch—hid from authority."

My pulse accelerated as I pressed against the Chinle and caught my first glimpse of those red cliffs etched between Utah's Henry Mountains and Colorado's La Sal Mountains. Few vistas in America could be as enthralling.

"This is a land of men with massive forearms who have always slept within the kiss of the most deadly pit vipers on our continent, each night, before they mine nuclear ore." As she whispered I discerned four distinct weather patterns.

My companion began to talk of youth and the only known inhabitants of the San Rafael Desert between her mountain, 7,000-foot Temple Mountain Emery County, Utah and Robber's Roost: Andrew J. Denney and his wife who dwelt east of the San Rafael Reef from 1922 until the mid-1950s.

"Rarely would Dad leave Temple and pass through this bioherm for a trip to our uranium property in the Slick Rock of Colorado, any of several uranium mills over one hundred miles away, or home to Price, Utah without stopping at Andy Denney's ranch house for companionship, education, hope, or a cool drink of water."

The brilliant desert morning began to play on my eyes as she recalled the first time she was overcome by a thundering herd of wild horses. Mustangs, exciting dust for miles, appeared from the west and completely surrounded their truck in the desert but one mile away from the ranch. Brenda had watched with terror and amazement.

Also, it was not far from Denney's Ranch that her father had pointed out wild buffalo in the mid-1940s.

On the south end of the ledge a cavity had been sculpt by wind, ice, rain, and 200 million years. Surrounded by piñon, this once served as her father's spa. The bowl was eight feet in diameter and four feet deep. Its water appeared clear and already warmed by intense sunlight.

"A frontiersman by the name of Swazy, the person reported to have discovered the first uranium, carnotite float in Wild Horse Canyon, at Temple Mountain in 1893, is given credit by Dad for telling him where to seek moisture on the San Rafael Swell. Mr. Swazy told my father of these 'Mormon Bathtubs' and to look for the shimmer of 'Cottonwoods' which occasionally signify a spring or pool of water is near. Domestic water is still over 25 miles away, really more like 40, unless you dare drink the Dirty Devil River!"

My rather glamorous guide looked directly through me as she continued.

"Was I incorrect when I told you in your gallery back in Laguna Beach that I 'had' a saga for you?

"The tale of uranium's growth is a story of Mother Earth first and foremost," feeling lasers upon my upturned face. "Did you know Gertrude Stein once admitted being able to rest only when looking directly into the most vibrant of suns?" fell on deaf ears. This lady did not have Stein on her mind, motioning me back down to the fissure of ascension.

Within the cloistered confine of our prototype Terra Van soothed by Waylon and Willie's plaintive lyric I watched in hypnagogic fashion as earths rare to our human experience passed.

"These are the very cliffs the Wild Bunch would squeeze through on their way to the Roost," Brenda mentioned above "the bright lights of Denver were shining like diamonds..." "Fewer human beings than you could ever imagine have passed under these mauve monoliths."

Martin Image, North Temple Wash, 1976

As we brushed Navajo Sandstone I recalled her telling of a 6-foot-high wall of water and boulders carrying her late father's vehicle over a mile back down this chasm as he clung to the sandstone.

"The San Rafael Desert, part of which we just crossed, still conceals the bleached bones of many who prayed for a blizzard's moisture before dehydrating, prostrate, upon this segment of their western experience."

Without a rudimentary appreciation of these very sacred environments one need not concern themselves with those, such as Brenda's remarkable father, who have slipped beneath our planet's crust in search of the rare earth called uranium and the energy from within its halo.

North Temple Wash and my mind were working as one, opening together onto a plain splitting the great reef.

"Watch for mule deer or desert bighorn sheep. Back in the early 1940s Dad used to spy elk wandering down from the Wasatch Mountains and crossing the Wasatch Plateau during the most extreme months of winter."

"For elk I guess it was like going to Palm Springs," I quipped.

The Terra Van suddenly hit bedrock and bounced close to a foot in the air jarring my flight of fancy concerning elk in Palm Springs. Fine, red, radioactive dust trailing through the wash gently overcame us as Brenda backed off bedrock, shifted into four-wheel drive, and climbed the steep bank of the wash.

"We'll search for the 'old road,'" she mentioned.

I was pinned to my seat.

With Dior sunglasses perched atop her head this little frontiersperson asked our vehicle to respond like her long deceased "Rocky Mountain Canary, Sally." We flew over the bank like a four-wheel drive ad and were soon upon an "old road" heading straight for Temple.

Our balloon tires and high suspension appeared to put us a stratum above the unique desert below, giving an eye-to-eye level view of rare earths which very few ever really studied. Ancient vanadium workings became the focus.

Again reading the environment as my mind this space-age pioneer exclaimed, "Vanadium, a chief byproduct of uranium, is a metallic element, primarily used to increase the tensile strength of steel. You will learn much more about it very soon.

"Look," she pointed to a weathered Wingate monolith suddenly visible to the west. "There she is!"

This was the moment I had been anxiously awaiting several rather intense months. My first substantial look at Temple Mountain, Utah, Brenda's "most beautiful mountain in the world," a landmark and environment of significance when the chronicle of uranium's mystical and industrial growth is researched.

Vernon Pick's "Hidden Splendor," Charles Steen's "Ma Vida," Howard Balsley's "Yellow Circle," Vanadium Corporation of America's Monument No. 1, on the Arizona side of Monument Valley, Lawrence

Migliaccio's "Vanadium Kings" on North Temple Mountain. All more than mere echo of a remarkable industrial evolving.

"You are now truly west of the largest reef in this world. Some have called this a land of too much life or not much other than bleached death."

Martin Image, North Temple Wash, 1976

Diors still in place, Brenda was rallying around 15- to 25-foot high fallen Chinle boulders. "Think of doing this in the opposite direction with close to forty tons of glowing uranium ore on your tail. The stories you will hear, I can tell you, about backtracking the San Rafael Desert and the Green River Deserts and on to the Atlas Mineral's epic mill in Moab, Utah or Vanadium Corporation of America's historic operation at Naturita, Colorado, stories of Blair Burwell, Denney Viles, Howard Balsley, are stories of strength and endurance and industrial history."

She was then silent for some time purposely trying to jar my fillings. It didn't matter at all. Suddenly, there was the entire Temple. On the floor or ceiling. That is one hell of a mountain, rather realm!

Our van slowed to a natural crawl, sort of Jurassic shuffle. Before us lay pure moonscape painted in rather mauve tones. Temple Mountain is the future and past in one blink. I was rather stunned. Before me, a mountain comprised of many known uranium-bearing formations, exposed, rising 7,000 feet above sea level, probably growing faster than it is weathering, is a very tight description. These include the Triassic System formation, the Lower Triassic Series Meenkopi Formation through the Shinarump or Mossback Formation then Chinle Formation and up into the Jurassic System.

We glided towards what appeared to be a wooden and very weathered colony of buildings, cabins, set against the base of the south side of North Temple, one "Beer Crate" and several Lodge Pole Pine facade cabins uninhabited by man for over twenty years, upon a moon!

As we stopped in the center of this abandoned uranium settlement the fine, red, radioactive dusts of North Temple Mountain again overcame us.

"If coming to study the past to truly understand what is necessary for new domestic energy potential to grow and become viable, you must remember that when my father and the others came to this land seeking uranium ore it was a new domestic energy potential and the heartrending story of its evolution is not too different from the story of all the struggling energy potentials here today." She was right. "What has happened, is happening!" Brenda sang above ". . . Sometimes it's heaven and sometimes it's hell, sometimes it's just in between."

"Remind me to take you to the Bureau of Mines Research Center and Oil Shale Pilot Plant in Reno, Nevada. Talk about an important domestic energy potential slow to develop! One reason is an apparent rivalry between the Bureau of Mines, which used to do all oil shale research, and the Energy Research and Development Agency, presently controlling the research program over oil shale."

"But uranium ore is what the public is interested in," I replied.

Martin Image, Well beaten path to the mines.

"When we climb to the mines you will catch the scent of industry and industrial pioneers. The story of uranium ore in the United States goes back long before the twentieth century!"

This is the kind of environment that one just had to touch for it must be remembered we trod earths scientifically rare to our earth. Thus it was not long before Brenda led me up another gradual bleached incline whose slope sheltered pockets of oxidized uranium and vanadium ores interfused with much petrified wood and pyrite shaded with aqua and orange sandstone laced with ancient desert pinyon.

Under these little natural epiphany scenes lay the Moenkopi formation. The first seven feet appeared to be sandstone and light greenish-gray and red-brown shale, very thin bedded. Suddenly, about one hundred and fifty feet of red-brown shale interbedded with numerous thin beds of ledge-forming sandstone and a few of almost ochre-colored shale and ripple, ocean, marked sandstone, 4,000 feet above the contact with sea level of a geological skeleton worthy of a Smithsonian all its own.

My guide's footsteps and the eternal droning of desert insect life were the only sounds.

This same exposed formation then introduced thirty feet of very sandy shale and red-brown shale followed by over forty feet of greenish-gray and limy-gray ripple-marked sandstone, almost laminated.

"No wonder it has been reported, not confirmed, that the Curies sent to Temple Mountain in 1898," Brenda whispered down to me. "I think have been told, we have her mine cart."

Another one hundred and seventy feet and layer upon layer of pink-gray fine-grained, thin bedded sandstone.

Uranium Seekers

Martin Image, View from entrance to Vanadium King No. 1, 1976

We were now well above the settlement and slowly tacking the path to the Migliaccio Vanadium King No. 1 Uranium Mine whose history was to unfold.

This Moenkopi still had over one hundred feet of strata of differing textures and colors and conglomerates yet to mark.

"Watch out for sidewinders, pygmy rattlesnakes, and flattened scorpions," Ms. Migliaccio whispered down interrupting my hyperventilating. "They are the color of the very earth you are on. There is no way I could get you to proper medical facilities within two hours. So please watch your every step and hand placement."

I had not even thought about snakes yet immediately felt a thousand cold, venomous eyes watching me, daring me to disturb their meditative slumber, to become careless for five quick seconds. I began tuning the sounds of my present, listening for the anticipated excited rattle.

Reaching my effervescing guide, "Tonight, when all is still inside the cabin, when the rats cease scurrying and collecting everything not tied down or boxed up, we will know the snakes are here. They hide in the cooler recesses of the old settlement buildings and just wait for the daffy pack rats to trip around." She was silent.

Scurry, patter, scuffle, silence!

"Your breathing."

Scurry, patter, scuffle, silence!

"The snakes just sit back for several days and digest their prey. Then all is still!"

She was looking to the northwest while speaking almost in a trance. Again, as if she had already heard too many scurry, patter, and scuffles a little to the west of the largest reef in this world.

"Five minutes and we will be at the portal No. 1; hurry your thinking and let's get on."

"Only if you tell about your father before the other uranium seekers," I called ahead.

"At the proper place and proper time," she replied. "It should be the portal of Dad's Vanadium King No. 1."

Ahead we would come face to face with the oldest uranium mines in any Shinarump or Mossback Chinle formation on the entire Colorado Plateau and few environments on the face of planet Earth have been more instrumental in the growth and development of mankind's search for domestic energy potential and domestic defense. The southwestern portion of Colorado, the southeastern sector of Utah along with the northeastern sector of Arizona and northwestern cut of New Mexico, comprise that silent dominion known as the Four Corners.

Uranium Seekers

Martin Image, Miners are people who make light
where once there was darkness . . . Royce

Few pioneer Americans have been more prone to listen for gentle murmuring from within their earth, or front the present with more humility, than persons who have been involved in the mining profession. Brenda assured me, "Deep within the soul of any mountain hundreds of millions of years old, or entwined with the blackness under any desert, miners listen!"

Miners are persons who make light where once there was only darkness.

A 50-degree blast of radioactive air cleansed me at the portal to the epic uranium mine. Uranium even smells different than other domestic energy possessing an aroma of our future.

Gazing to Utah's isolated Sinbad Country Brenda continued, "A reporter from the old <u>Life</u> magazine was killed flying to talk with my father; his plane went down on the way over here from Moab, Utah or Grand Junction, Colorado, I can't remember which. I have something in my hiking bag. A little piece done on him in 1954. Reading it here in the portal to No. 1, the first time you have ever been upon Temple, being able to glance up and see the Sinbad may not be too bad a place for you to meet my father for the first time," spoke the heiress directly to me. "You should meet him here and in silence." All was certainly still at that moment.

She handed me Henry W. Hough's periodical which served as the official spokes vehicle of those associating with America's uranium industry from 1950 until the mid-1950s. Its pages yellowed and soiled by hydraulic oil stared back at me. There was nothing else to do but read:

As some of the pictures with this report shows old vanadium workings still pock-mark the cliffs in the Temple Mountain area, about 35 miles southwest of Green River.

Many operators have tried their hand in the area, which is one of the state's principal producers today. AEC carries on constant drilling programs here . . .

Vernon Pick's Alpha (first) mine is here, and some day when his Delta (fourth) bonanza 20 miles southwest of Temple Mountain proves less engrossing, he may come back to the Alpha.

The air and vista were more rewarding than this. Historically, however, Vernon Pick was always good copy.

K. S. Mittry's new Uranium Industries, Inc., of Grand Junction is carrying on its production at some of the Lawrence Migliaccio claims at Temple Mountain, formerly operated in the name of K. S. Mittry Construction Company of Fort Collins, Colo . . .

Other claims of Mr. Migliaccio have been involved in a lengthy hassle with E. C. Frawley's Consolidated Uranium Mines, Inc. Mr. Mig., a husky one-time beer wholesaler at Price, took a fortune in vanadium ore from Temple Mountain between 1941 and 1948, when he met Mr. Frawley.

"So there is a fortune associated with Temple Mountain, in print, just as I had been told by Brenda," I said out loud, trembling at the thought of men surely vying for this very portal. Suddenly becoming one small person this saga of America's West unfolded further:

While lawyers battle and skim off the gravy, Consolidated operates the disputed claims along with its other properties pending an eventual settlement...

At Temple Mountain, operations for Consolidated are supervised by J. G. Moulton, a young mining engineer who worked at Cripple Creek with "Cresson" and for the "Vanadium Corporation of America" at Naturita...

Directing it all from his offices in Salt Lake City's Darling Building is E. G. Frawley, President of Consolidated. He is an attorney who had engaged in various kinds of metal mining operations for the past 30 years, all over the United States, and in Alaska, Mexico and Panama...

Newspaper article graphic.

Frawley has had his battles and has received some wounds, notably from Lawrence Migliaccio of Price who we mentioned earlier. Today, Mr. Frawley is sitting pretty with a capable staff including a blonde office assistant who is as intelligent as she is beautiful.

"Litigation slowed most uranium pioneers to a crawl, at times Dad was an honest man to his very bone and saw that everything, every little problem was handled according to the letter of the law." This young lady continued, "Two infamous Judges were the ones he always faced, one was the late Willis Ritter of Salt Lake City."

Even I had heard of Ritter on "60 Minutes" recently.

"Hush up for one minute," I replied, nodding towards the page once again:

. . . life should continue to be real good for Mr. Frawley and his stockholders—if he can continue to discover and block out more uranium deposits, if he can keep out of Mr. Migliaccio's way and if that luscious blonde sticks around to help him run his neat little empire.

The magazine was not closed before Brenda began,

"You just try the same trip you experienced this morning without a twenty-thousand-dollar self-contained unit, put on top of that the year is 1893 or 1950, it does not matter much. Nothing was or is constant. God, they were brave men.

"That is what I am ready to hear about," I quietly replied. "I am touched by your father's presence. These are epic earths. But we will need to chronicle others, those before he first touched this energy potential here."

Brenda began to interrupt and I stopped her short. Being the only corporeal male on this one-hundred-million-dollar mountain at the moment, I felt rather supreme.

"Dad was into oil on the Unita Plateau in northern Utah and down by Mexican Hat before he came to this property."

"I don't give a damn about natural gas and oil right now. We can only interest ourselves in the newest domestic energy potentials," kicking over a carbide can. "The public needs to know what's coming and that still boils down to uranium power, natural gas, oil shale, some geothermal, a little solar and lot of coal."

"All but uranium and coal will turn out to be just footnotes in history," my guide exclaimed. "And I recently read that nuclear wastes are approximately five million times smaller by weight than those produced by coal-fired power plants. With the technology to handle the nuclear wastes further advanced, let's get on down to the settlement, Mr. Poet."

It was soon evident this realm springs to life after each laser retreats, as Brenda prophesied. However, at this moment I sat nursing minor wounds from the day's activity while observing sunset.

The chill winds which drove us from Vanadium King No. 1 abated. My guide moved our black vehicle, now baked red with an entire new coat of radioactive dust and sand, close to the weathered cabin erected from discarded beer crates hauled to the settlement in 1942 by Lawrence. She set this skeleton of her past up as sleeping quarters.

Looking west towards Temple Mountain's geological crown accentuated by the rapidly darkening sky reminded me of the famous Paramount logo. Weary, I momentarily dozed.

This nod was suddenly interrupted. My eyes popped open, I swallowed hard, pulse quickened, and in my ears reverberated resounding similarities to firearms discharge.

Focusing, this alarm was merely intensity escaping ancient mesquite being incinerated by Ms. Migliaccio. Apparently my little nod had run on.

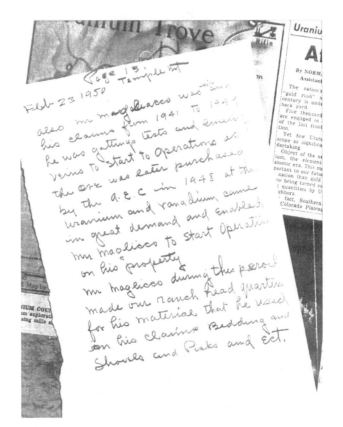

Andy Denny letter graphic

"Well, frontiersman, this desert may let one rest on the first night of your first trip but you best become acclimated to having a fire collected and ready to ignite before nighttime." Her exotic beauty was intensified by natural firelight. "I have seen a half a foot of snow fall in thirty minutes on this spot and one never wants to probe deadly desert earths in the dark, for light! Those slight nicks you received during our little climbs have already made you aware nuclear earths bear attention!"

I nodded agreement and sat to enjoy a true Pinyon fired Colorado elk steak.

"A person could sleep with vipers taking a shot at a few hundred million dollars on a steady diet of this."

"You must be willing to go on when the steak turns into beans and the beans almost to air, if you ever plan to take a hundred million out of here," retort the heiress. Her eyes blacker than the soul of her mountain this night!

The rocks gathered to build the pit glowed from contact with intense heat.

"Don't touch those," my guide warned. "These rare earths are the force to draw humanity for the past 4,000 years, at least!"

It would be remiss not to mention our transistor radio tuned to KSTR in Grand Junction, Colorado did offer, ". . . And there is nothing cold as ashes, after the fire's gone out . . ."

"The Uranium Magazine article was inspiring all right. Recite to me a lyrical history of this energy industry."

"There is no lyric which compares with one's present on this desert," Brenda snapped back. "However, Mr. Howard Wilson Balsley of Moab, Utah is closest to metaphor when uranium's actual history is researched."

She ceased a moment. The stars were out. From our camp located five thousand feet above sea level the nearest man-made light twinkled twenty-five miles away in Hanksville, Utah and over the two-thousand-foot sawtooth reef to our east. That is a lot of sky on a cold and clear evening.

T. Protopappas Migliaccio Cabin graphic.

"So you want a little account of uranium's industrial growth in America, do you?" beamed my companion. "It is a story Father used to love to tell, and on this very spot!"

"We shall go see Mr. Balsley soon, but you start."

Temple Mountain's great stone face, illumined by moonlight, appeared as a pockmarked monolith cave-like openings with long trails of black slag spilling from each, like tears. Not merely its size but an energy from this mountain seemed to laugh at me.

"Andy Denney, whose ranch they call the Garvin Ranch . . ."

ORAL HISTORY AND OUTLINE MATERIAL

Martin Image, Electric Desert, 1976

The southern Utah desert conceals its treasure and has concealed it from all but the keenest visions, since the creation. It is a land of extremes. Most of its realm suffers miserably from the gasping attempts at supporting too much life and so soon returns to the bleached white death from whence it amazingly sprang. The trickle that falls between this life and this Death moves always vague and ghostlike upon the expanse. Those who have ventured its washes, mountains, dried creek beds and brush islands never find things quite the same as when they left them last. Some have called the heart of this desert the Goblin Valley.

Just north of the valley, hidden by the San Rafael Reef, stands a 6773 foot monument to creation. Those close to the deserts heart, in the Goblin Valley, call this earthrending THE TEMPLE. The handful that has lived on it, in it, and above it and crawled beneath it and through fifty miles of tunnel have called it Temple Mountain. It is in Emery

County, Utah. The Temple can now be found on most maps but was once maybe only felt as an unseen sensation throughout the entire world. This small touch that is felt from The Temple has left scars on mankind which even today cause strong men to move as vague gestures only across their expanses.

A United States Government survey dated December 2, 1912, reads in part:

> Honorable Thomas Hull
> U. S. Sur. General for Utah
>
> Dear Sir:
>
> In the survey of the Orinoco, etc., lodes embraced in Mineral Survey U. S. 6240, surveyed under instructions dated Nov. 4, 1912, it became necessary for me to establish a U. S. Mineral Monument as there was no approved land corner indicated within two miles of this survey. On Nov. 10, 1912, I set a sandstone 40 in. long, 20 in. wide and 6 in. thick, 20 in. in a mount of stone and built a conical around of stone 6 ft base, 4 ft high, northwest of the monument 5 ft. distant. I chiseled U. S. M. M. 246 upon the stone and a x to indicate the exact point. A point on a sandstone ledge marked B. R. x U. S. M. M. 246 bears S. 81° 48' E. 33.7 ft. A point on a sandstone ledge, marked B.R. x U. S. M. M. 246 bears N. 39° 20' E. 10.8 ft. A point on a sandstone ledge marked B. R. U. S. M. M. 246 bears S. 7° 45' W., 79.4 ft . . .
>
> Uranium's chronicle embodies the truths that were realized in the fortunes of those who have come in close contact with United States M. M. 246x. All who have stood upon TEMPLE MOUNTAIN, or within sight of it, have stood within the boundaries of that mineral monument designated as U. S. 246x.

Those wishing to visit this region should take note of the following directions, given by one who for a great time labored in that land. Your guide in THE FORTUNE OF TEMPLE MOUNTAIN then, shall be, in part, the voice of Lawrence Migliaccio and all miners of this earth.

NOTICE IS HEREBY GIVEN, THAT the undersigned, having complied with the requirements of SECTION 2324 of the REVISED STATUTES of the United States, and the Local Laws, Customs and Regulations of this District, have located Fifteen hundred feet in length by 600 feet in width, on this the Vanadium King, Lode, Vein or Deposit, bearing Gold, Silver, Copper, Lead and other valuable minerals situated approximately half mile Easterly of U. S. M. M. No. 246 which is about 46 miles Westerly of Green River Utah in the Temple Mountain Mining District, Emery County, Utah . . . State of . . .

The Temple is an energy. It is a mountain of uranium that is pregnant with what are perhaps the most unique deposits of the precious stuff to be found in the world today.

Al Szabo Image, Twilight at Temple Mountain

Hues of yellow and black continually ooze from within Temple, changing as if iridescent, never ceasing in charting new directions for the mountain's design.

A 1957 Bureau of Mines Information Circular from the Library of Congress defines the general geology of the Temple as follows:

Temple Mountain lies at the southeast flank of the San Rafael Swell. The swell is a great domal uplift about 100 miles long and 50 miles wide, with its main axis striking northeast. The core of the swell is enclosed in a steep rampart of sandstone cliffs encircling the highly eroded Sinbad Country...

Formations exposed in the immediate area range from the Permian Coconino sandstone to the Jurassic Navajo sandstone. The Coconino sandstone is ordinarily cross bedded, gray, and fine to medium grained. Where it is exposed to the north in Black Box Canyon of the San Rafael River, it is approximately 700 feet thick. The bulk of the overlying Permian Kaibab limestone is dark gray to black, is often sandy, and contains chert concretions with calcite and asphalt cores in the upper part. Ninety feet of this marine limestone has been measured 13 miles south of Temple Mountain. Overlaying the Kaibab limestone is the Triasaic Moenkopi formation. The Moenkopi is approximately 550 feet thick north of Temple Mountain and is composed chiefly of red to brown shaly siltstone and lenticular beds of fine-grained sandstone... Overlying the Moenkopi is the Triassic Chinle formation... The Triassic Wingate sandstone occurs above the Chinle formation and averages about 360 feet in the Temple Mountain area... The tower like peaks of Temple Mountain are isolated remnants of the Wingate sandstone... The Jurassic Kayenta formation lies above the Wingate...

The area is crossed by a series of faults, which generally strike east-west and are nearly vertical. The vertical throw of these faults ranges from several feet to several tens of feet, with various displacements along their strike. A number of irregular collapse features in the area have complicated the structural geology and have caused odd-shaped features. Bleaching and alteration of the rocks are widespread, with the areas near the collapse features most affected. The principle ore deposits occur in the Chinle formation, although uranium-vanadium mineralization has been found in other formations in the area.

The Sinbad County is a hermit land . . . alone. The history that surrounds the rock, together with the stark vastness that surrounds the whole legend, diminished everything in this presence.

As the eye moves upwards through 6773 feet of rock, so ageless, that eyes long, ascending journey is punctuated just before the sky by the Kayenta Crown, set in such a way that it seems almost unnatural, as if purposefully placed amid the chaos of a world in boiling infancy.

Martin Image, Balanced Rock, North Temple Mountain Wash, 1976

Looking Southeast from the portal of number six uranium mine, otherwise known as the Migliaccio Vanadium King, one sights, through a fault in the massive San Rafael Reef, the environment of Robbers' Roost some sixty miles off. The hold-up where Butch Cassidy and the Sundance Kid and many of the other legends of the American West hid from all but the eyes of eagles. It was here that these swarthy men, the end of a romance in America's past, escaped the law by passing through the Great Reef at Temple Mountain and crossing an arid expanse to the cliffs of Robbers' Roost. Only the bravest of pioneer spirit or the most desperate ever set upon the Sinbad Country. The Bureau of Mines Circular 8711 describes the environs as follows:

The Temple Mountain district is 35 miles southwest of Green River, Utah, and is just within the eastern escarpment of the San Rafael Reef. The San Rafael Reef is a roughly elliptical belt of massive, tilted sedimentary rocks encircling the heart of the San Rafael Swell. A highly eroded terrain approximately 40 miles long and 15 miles wide within the area bounded by the reef known as the Sinbad Country. Temple Mountain is a prominent and conspicuous landmark. The San Rafael Desert lies east of the reef . . .

Calyx Flat is bounded on the east by the Wingate and Chinle cliffs, which are part of the reef, and on the northwest by island like remnants of the Wingate and Chinle formations, which form Temple Mountain. North Temple Mountain rises to 6,773 feet above sea level and South Temple Mountain to 6,579 feet.

The sedimentary rocks of the San Rafael Swell have been carved by erosion into a topography that is rugged and picturesque. The more resistant sandstones that form the eastern reef rear suddenly upward as high as 1,800 feet above the San Rafael Desert and are tilted as steeply as 85°. The eastern rampart of the reef has been incised by numerous and impressive washes and . . . Immense erosion has cut the Sinbad Country, leaving mesas, valleys, flats, and canyon lands in random arrangement.

Most of the Sinbad Country is barren of vegetation, except for sage, cacti, and associated growth, although some pinon and juniper grow in the higher areas . . .

The mine roads are often muddy and slippery in winter and covered with a thick mantel of dust in summer. There are no perennial streams in the immediate . . . Water for domestic use is trucked from Green River.

The nearest rail point is on the Denver and Rio Grande Western Railroad at Green River, 43 miles from the camp.

The scope of our story must be getting a bit larger . . . considered virgin, but also with 50 million . . . of ore that has left through the washes and to the west. Most of Temple's history occurred when water was over . . . The man who then labored on and within THE TEMPLE have learned the true joy of moistened lips.

Traveling those 43 miles to the nearest water often took these men from one and a half to two days, just to reach the warm cup. The desert showed little mercy for those who tried to cross it. Even today the small ribbon of asphalt that runs from the Green River Junction to the Temple Mountain Junction often vanishes, from even the most piercing gaze, quilted by the fury of a blizzard or sandstorm.

It is reported that pioneers of the Swell were was upon the Sinbad as early as 1935 in search of uranium they felt would take mankind to the Galaxies and give warmth to the multitudes with a flick of a switch.

It should be noted at this time that the United States Atomic Energy Commission was set up in 1946 and its administrations have since tended to bend a little towards commercial interests. Its initiation was an attempt to stockpile as much uranium as possible for as little capital outlay as possible. However, there were cash bonuses involved for all participants. The American public should also always keep in mind that the persons who supplied the United States Government with the Uranium used for the Atomic Bombs that melted the people of Japan, had no idea that the Government was eating a war device.

THE FORTUNE OF TEMPLE MOUNTAIN is a story that begins long before its human participants arrived upon the scene. Before Madam Curie; before Charles Steen opened mines that have shipped a reported $150,000,000.00 worth of ore (and Steen still wanders alone upon the western lands in search of another lode); before the Internal Revenue Service auctioned off his mineral collection in 1975; Temple's fortune stood before all this.

W. L. Dare, mining engineer for the Department of Interior, discusses the early history and production of Temple Mountain:

"Available information indicates that the Temple Mountain deposits were discovered in 1903. Prospecting and development work was carried on in the area from 1904 to 1913. Mining was begun in 1914. The mines produced an unknown tonnage of high-grade uranium ore between 1914 and 1921. A small tonnage was produced in 1941 and 1942.

Lawrence Migliaccio, Vanadium King No. 6 Mine.

J. M. Boutwell (in a 1904 Geological Survey Bulletin) tells of the discovery of "considerable" deposits of a black, vanadiferous sandstone and some carontite float in "Wildhorse" canyon by sheep herders in the fall of 1903. He adds that these discoveries were later prospected in dense, carbonaceous sandstone which contained combustible matter. After burning, it yielded a residue that contained vanadium. Boutwell never visited the discoveries but describes them as southeast of the San Rafael Swell and writes that the deposits were reported to occur in large quantities of a "Cretaceous" formation.

Wild Horse Creek empties into the Muddy River about 15 miles southwest of Temple Mountain. Upstream from the junction the creek passes between the Morrison capped East and Middle Wild Horse Mesas and then skirts the eastern escarpment of the San Rafael Reef. It splits about 2 miles south of Temple Mountain, with the left form turning west and cutting through the reef as Wild Horse Creek. The other split continues an additional mile before it enters the reef as South Temple Wash. From Boutwell's brief description of the ore, it is believed that the discoveries he mentions were in the Temple Mountain area, for it is conceivable that the split now called South Temple Wash could have been known earlier as Wild Horse Canyon. The numerous old workings along the outcrop of the Chile formation in South Temple Wash could have been known earlier as Wild Horse Canyon. The numerous old workings along the outcrop of the Chile formation in South Temple Wash indicate that the initial discoveries of oxidized ores were easily accessible from the canyon. Lawrence Migliaccio of Price, Utah, said that old-timers in the area reported that a frontiersman and stockman named Swazy found the first ore at Temple Mountain about the turn of the century."

THE FORTUNE OF TEMPLE MOUNTAIN is a chronicle of frontiersmen who burned strange ores, upon the Great Domal Uplift of the San Rafael Swell as two centuries of American history passed each other in the dusk.

Dare continues to discuss the very early geological and commercial history of THE TEMPLE.

"In 1914 F. L. Hess reported:

> During the year the commercial production of uvanite, a radium bearing mineral new to science, accompanied by other uranium minerals, one or more of which are yet to be described, was begun at Temple Rock, 45 miles southwest of Green River.

Hess wrote in 1917:

> The Chemical Products Co. of Denver started work during the year on the uvanite ores from Temple Mountain and had radium salts ready for crystallization.

Continuing, in 1918, he said:

> The Chemical Products Co. of Denver, continued to isolate radium front uvanite obtained from Temple Rock and bought the claims.

The Chemical Products Co. sold its interests at "Temple Rock" to the Ore Products Corp. in 1919 . . . in 1922 Hess reported . . . all production before 1922 was from Temple Mountain." In the same article Hess said:

> "When the radioactive deposits at Temple Mountain were discovered, it was not the asphaltite that attracted attention, but its oxidation products, for oxidation gives rise to a remarkable series of bright-colored uranium and vanadium minerals that would attract any prospector's attention."
>
> "The earliest production of uranium-vanadium ores from the Colorado Plateau was sold for extraction of their radium content. Beginning at the turn of the century, vanadium was used in small amounts for experimentation in steel alloying, most of this work was being conducted in Europe. Uranium was also tried for steel alloying, most of this in conjunction with vanadium, but the coloring of glass and pottery products was the principle use of the yellow uranium minerals. Before the first World War, European municipalities and hospitals were the principal buyers of uranium and radium products in the United States. The demand from Europe fell when war started, for European

money was channeled to other markets. There was virtually no market for uranium and radium ores produced in this country in 1915. However, the demand for the byproduct vanadium increased slowly as development of alloy steels progressed. During this period the radium-based material for watch faces was developed and became the principal use for radium in 1917. Production in the Temple Mountain district continued until 1921, when unfavorable business conditions both in Europe and in this country greatly curtailed the market.

L. Migliaccio reported that he, John Davis from Price, and John Adams from Green River mined high grade oxidized pods exposed along the base of the Wingate sandstone at the northeast end of North Temple Mountain during the summer of 1941 and spring of 1942. The ore was hand-drilled, sacked, and packed off the mountain by the men to mules below. As a result of this work 6 tons of high grade ore was sold to the mill at Naturita in 1942."

The only published use of high grade uranium ore by the United States Government in the first part of the 1940's was to link THE GREAT CHAIN above the islands of Japan.

Lawrence Migliaccio, North Temple Wash

There are very few industries whose mention can cause the mind to wander or the pulse to quicken when they appear in prose or dialogue as does mining.

This is attributed to miners. It introduces a few of the most famous and daring miners of the twentieth century. This outline shall cover the technique used by Lawrence Migliaccio, Charles Steen, Jack Turner, The Atomic Energy Commission, Leo Cline and the Boys from Cripple Creek, Colorado, Howard Balsley, Andrew Denney, and the Swazy Boys . . . frontiersmen in every sense of the word.

In fact, in order for the reader to glean the complete story of the development of the atomic age and reach some sort of overview, they must first be made well-versed in uranium and gold mining.

A 1967 Utah Mining Association Bulletin discusses the Historical Development of Utah's Mining Industry.

The history of Utah's growth from a desert wilderness to a thriving populous commonwealth is a story of determined men subduing the earth and learning to use its treasures. The earliest record of knowledge and interest in the minerals of Utah is noted in THE GREAT SALT LAKE by Dale Morgan. Mountain men who had trapped beaver in the valleys and streams around Great Salt Lake and Utah Lake told a group of Mormon pioneers wintering in 1846 in Pueblo, Colorado, that ". . . in a ridge of mountains running through the lake large quantities of precious minerals were found."

The Mormon influence in the Uranium Industry of America is and has been great. That is why the author has engaged the aid of prominent Mormon teachers and historians from the southern portion of Utah.

Martin Image, Miners Cabin, Temple Mountain, 1976

"Petroleum-type organic material is associated with the uranium deposits at Temple Mountain. It occurs along faults and fractures, bedding planes, lithologic contacts, and impregnates permeable strata... The uranium-bearing material may be a friable, dull-black, greasy petroleum, which forms bands along bedding planes, fills fractures, or occurs along faults, or it may be a hard, friable, dark-gray or black material, which forms pods, pellets, and streaks along bedding planes and fractures. Petroliferous material, which contains no uranium, may occur as a semifluid or fluid, tarlike substance. In the fluid form this substance can be seen oozing from the walls at various places..."

The author has already mentioned the fact that hues of yellow and black continuously ooze from within Temple, changing as if iridescent, never ceasing in charting new directions for the mountain's design. Dare continues:

"Much of the semifluid tar impregnates the porous sandstone and imparts a pronounced greasy feel to the rock. Broken particles of this impregnated sandstone seem to "creep" down a muckpile and will stain the fingers. The non-uranium-bearing petroliferous mater may also occur as brittle, black pellets or streaks. The asphalite material releases a fetid petroleum odor when first broken. This is especially noticeable in stopes where high-grade ore is being mined.

The asphaltic-ore bodies in the Moss Back member of the Chinle formation are usually found in the more massive sandstones. The conglomerates are also mineralized when within this main sandstone stratium. The ore occurs in a series of closely spaced, high grade pockets surrounded by halos of low-grade ore, which sometimes connect the high-grade pods. The ore varies in thickness following the beds in general, although not in detail. The ore in places appears to be controlled by structural features..."

We have begun our sojourn to the throbbing heart of THE TEMPLE, totally controlled by structural features and in search of "halos" in the darkness just a little above the GREAT DOMAL UPLIFT. Leo Cline, the leader of the boys from Cripple Creek, once spent an eternity in the darkness just a little above the Great Domal Uplift, led astray by one of those small "halos." His story shall be heard shortly in our outline.

"At the Calyx No. 8 mine (mined by Leo Cline and the boys from Cripple Creek) ore has been found to follow along north-west striking fractures. Most of the ore is present at the base of the Moss Back near its contact with the lower siltstones and mudstones. However, ore has been found occasionally on an upper level which is 20 to 30 feet above the contact. Occasionally, the ore pods overlap, and sometimes there is a noticeable split in the ore sandstone with the ore occurring above and below barren mudstone. The ore at one point in the Calyx No. 8 mine seemed to roll sharply upward to form an ore pod some 20 feet thick...

Martin Image, Tracks find daylight. 1976

The unoxidized asphaltic ores mined at Temple Mountain are reported to contain uraninite within the petroleum mass . . ."

The boys from Cripple Creek often found these pods by entering Temple Mountain within the witness of United States Mineral Monument No. 246x:

"Both mines are developed by small-diameter circular shafts, sunk with a truck mounted, 36-inch-diameter calyx drill. The Calyx No. 3 shaft is 117 feet deep, with the station level at 105 feet . . . Both shafts are lined with 12 gage sheet steel for a short depth below their collars to prevent caving of the less consolidated surface material. The shafts were sunk, for the most part, through sandstone. The walls of the shafts are smooth and stand well. Small amounts of rock have broken out of the walls where minor lenses of shale were penetrated.

CALYX FLAT is dotted with many small calyx shafts. It was the belief of Consolidated Uranium Mines, Inc. that this relatively cheap method of developing the individual ore bodies eliminated the need of crosscutting from one ore body to another, minimized haulage distances, and provided, a unit operation for each lessee. To initiate production, the ore bodies were not completely delineated when the shafts were drilled. It later developed that some of the shafts could have been located to better advantage relative to the ore bodies or to underground haulage grades . . .

Both operating shafts were sunk through the ore, and ore pillars were left at the stations . . . the man-trip opening from the shaft to the station is a slot through the shaft pillar about six feet high and two feet wide, with the loading chute on the opposite side . . ."

A Tonopah, Nevada newspaper, dated Friday, February 3, 1956, contains a front page article which reads as follows:

"Undanted by snow and bitter cold, Jack Turner and associates on Monday pushed off a vast exploratory and development program that has as its ultimate goal the opening of a major uranium field in the Tonopah district of Esmeralda county. The group has a long term lease on 52 patented claims owned by the Lambertucci brothers.

Grinding out test holes to varying depths is a huge rotary drilling rig operated under the watchful eyes of Turner's uranium engineer Joe Dowd . . .

Some indication of the size and efficiency of the immense drilling rig is the fact that on-lookers clocked it on Monday punching out a 140 foot test hold in two hours and ten minutes. The rig can go to a depth of 1000 feet, it is said . . ."

Uranium Seekers

Martin Image, Headstone for brother miners world wide. 1976

It is clear that ours is a story of the Uranium Mining industry at least from the time when men climbed a seven thousand foot living mountain of uranium to hand-chip and drill ore to be carried down that seven thousand feet on their backs to their mules below ... "Grinding out test holes to varying depths ... (with) ... huge rotary drilling rig operated under the watchful eyes of Turner's uranium engineer Joe Dowd..."

It is a document mentioning The Swazy Boys ... frontiersmen, of Leo Cline and the Boys from Cripple Creek descending into the earth in tubes two feet in width in search of halos of uranium, of Lawrence Migliaccio searching The Temple with one stick of dynamite at a time as the great rotary drill punched that earth a thousand times.

"The ore is mined by open stoping, with random pillar support. The irregularity of the ore is illustrated by the labyrinth like plan of the underground mine workings. The ore pods sometime roll sharply and often appear to be controlled by fractures and by the mudstone and siltstone contacts. The ore often splits into two separate horizons, with waste between them. Surface drilling ahead of the existing workings is important for indications of additional ore, but equally important are the skill and judgment of the miners to recognize the sometimes erratic ore trends.

Pillars do not follow any particular pattern, for their spacing and size depend on the width of the span and the nature of the overlaying rock. Irregular pillars of waste are left for support where waste islands are encountered within the larger ore bodies. Ore pillars are left sparingly. The rock overlying the ore consists of barren conglomerate or mudstone and sometimes barren or low-grade sandstone. The mudstone sloughs easily. The lenses of conglomerate, for the most part, stand fairly well, although slabs will break loose if the contact with an overlying mudstone exposed . . . (a slab may weigh several tons)."

Uranium Seekers

Martin Image, (cover image "Hand which has labored for humanity."
Frank Migliaccio, Green River, Utah, 1976

Frank Migliaccio, of Green River, Utah, was Lawrence Migliaccio's brother and cement finishing foreman on both the Boulder Dam and the Grand Cooley Dam. He is one who has witnessed the finishing of many foundations in America a grain of silica at a time. He was upon the crests of Temple with his brother, Lawrence, long before the foundation of the Atomic Energy Commission.

The single most remarkable thing about this man is the work of his HANDS. They are the hands of the man responsible for over 20,000,000 tons of concrete being poured to light and power all the Pacific Northwest and major portions of America's Southwest. All twenty million tons were poured before the World War II.

Many times the government of the United States has attempted to give him a job guaranteed for life; and just as many times Frank Migliaccio has turned them down.

When a man this close to the earth and this understanding of her completeness speaks, it is time that others finally listen.

"... I don't know exactly when I quit or when I started ... I was cement finishing foreman for the whole Grand Cooley Dam ... it was the largest in the world ... Boulder Dam, I was foreman of that too ... I started sometime in 34' for five or six months and the boss left and went to the Grand Cooley Dam ... they were supposed to send a boss over from the day shift to the night shift and the foreman of the night shift told them to take me and put me to work ... I'll never forget it ... and I was boss for about 11 years ... there were 11,500,000 yards of concrete in Grand Cooley ... after Boulder Dam was completed I went up to Cooley ... boy at that time they had lines for the jobs and if you didn't live in Washington, you couldn't get a job ... the personnel manager asked me ... "Bud, what are you doing here?" I said that I was looking for a job and he said, "Get behind the line" ... I asked if Laughton Carey or Johnny Tacky were there ... he said, "Come on in, I want to talk to you" ... They said they couldn't give me a cement job because there were state men enough to hire ... but ... if I would hire on in structural steel, they would transfer me over ... two weeks later they put me on as foreman ... it took seven years to finish that job ... there were two phases; they built the foundation and then the "high dam" as they called it ... different contractors.

The bottom was four contractors and the top dam was comprised of eleven companies ... they called them all CONSOLIDATED BUILDERS, INC.; I still have my hard hat ... 119 men died at Boulder; they said that when we use to strip the forms there were hands and legs all over in the concrete ... its a lot of bunk ... I even heard it while I was working there ... hell, there was an inspector for every ounce of concrete they poured ... if there was a stick in there, we would stop the pouring and pick it out ... 96 men got killed at Grand Cooley ... but it did take seven years ... you take a city of 7,000 people and not many died on Cooley ... it was dangerous, sure ... we tried to form a union at Boulder ... I gave them five dollars, but we couldn't organize ...

Those dams are there forever ... they are fixed for earthquakes too ... they are keyed and poured in fifty foot blocks ... if there is any slippage, they tighten up ... grout pipes for when the dam cools and opens up and cracks ... they pump in cement and water. They pumped all the rocks and crevices around these dams for seven years with cement and water ... filling all those cracks ...

I was general foreman on the Friant Dam out by Fresno ... I was general foreman for this Geneva Steel plant up here ... that's the largest plant in the WEST ..."

Far up Shitamaring Creek Canyon near Ticaboo, Garfield County, Utah in the heart of America's frontier west one gets the faint impression, as the second shift from the Plateau Resources uranium mine comes in to eat, that Frank Migliaccio most certainly has been "messing around" in our American West.

Martin Image, Frank Migliaccio, Green River, Utah 1976

". . . they offered me a superintendent job down in Savannah, Georgia at the Savannah River Nuclear Power Plant. I went down there during one of those strikes . . . just to look around and filled out an application . . . the secretary told me to stay right there after looking at it . . . she said that she as going to get the head superintendent . . . that was a big job . . . a hydrogen bomb plant . . . they brought the superintendent out . . . and he said that they couldn't place me as a cement finisher, but that I could be superintendent of laborers . . . I may be back I said . . .

Frank Migliaccio is a testament to all persons of the earth who have calloused hands in laboring for humanity.

After he completed the Grand Cooley Dam he returned to Green River and the San Rafael Swell and read the Bible for most of seven years, simply as a well-deserved rest.

Another Temple Mountain miner, Jack Turner, set upon the "Sinbad" by invitation from Migliaccio; built his home and family in the face of the "Sinbad's" fury a decades ago, the windows of which still stand today. They stand upon the San Rafael Swell for you to peer through. The "New York Post" told the world that this man was Jack Turner, . . . (one half of the) . . . RAGS TO RICHES PAIR . . . (who turned) . . . URANIUM INTO GOLD. Three years ago, Jack and Agnes Turner stopped smoking because they couldn't afford to buy cigarettes. Now they talk about bucking the BIG THREE in the uranium market.

The BIG THREE are the United States Vanadium Co., The Vanadium Corp. of America and Climax Uranium. "Jack-the-Giant-killer Turner, 36, is bent on preventing a uranium monopoly. He also would like a few more millions for his own family. "Some say I've made $5,000,000 in three years," he once stated. "I haven't had a chance to figure it out. But it's safe to say that uranium from my mine was in the bomb that fell on Hiroshima and that pretty soon there will be a uranium explosion in the financial world."

Turner was interviewed before he left for the American Mining Congress at San Francisco.

When the A-Bomb burst, over Hiroshima, Turner was busy dealing with buzz bombs as a fire control technician with the 407th Anti-Aircraft Battalion in Europe.

Turner's great grandfather was born in Moab, Utah, and Turner's father staked out the first Utah uranium claim, in 1896. Directly from high school, Jack Turner himself entered the ore business.

The classmate he married drove his ore truck. In youthful optimism, they hoped to make a fair living out of uranium. But when the war was over they saw no future in it and moved to California. Then they heard about the government ore-buying program. They returned to Moab. The Turner's income doubled, year by year Turner acquired hundreds of 20 acre claims in the area . . . He showed Charlie Steen where to look for his first million and now Steen has claims worth $150,000,000 . . .

Still Turner was about to give up three years ago, after mortgaging his home, pawning all possessions, and scraping up $15,000 for an Atomic Energy Commission lease. Had his wife's family not staked him to a week's worth of groceries, he might have sold a half interest in his mine for $2,000.

On the seventh day he hit pay dirt and soon the AEC surveyors estimated that he had more than $2,000,000 worth of ore in that mine...

He said:

"I started with pick and shovel and I have either luck or a knack at finding uranium and I'll do what I can to prevent a monopoly in this ore which may determine the future of the world." (Wednesday, September 22, 1954, New York Post).

H. D. Quigg, a United Press Staff Correspondent ran this article on Jack Turner in June of 1955:

Jack Turner, a mud fence-plain, leather touch citizen from Utah who grew up among cattle, sheep, cactus, and-acres of throwaway dirt containing a worthless mineral called uranium, is in town with a look of misery on his face.

He is struggling under a monstrous burden: $21,000,000 and it's growing bigger all the time. Being a multi-millionaire can weigh a fellow down... like a yoke of solid gold...

Walter Winchell, in the Daily Mirror, and his Man-About-Town column called Jack Turner:

Man About Town

Most lavish tipper in town is Jack Turner, 36, of Moab, Utah. $50 and $100 tips.

Cholly Knickerbocker says in the New York Journal-American, Friday, October 15, 1954:

Fashion commentator Ethel Thorsen entertaining Jack Turner, the uranium millionaire, and Lee Pelzman, the plastics man at the Colony... Another uranium millionaire, 36 year old Jack Turner, lavishingly tipping while in Gotham...

Uranium Seekers

When Jack Turner left the red cliffs of Moab and went to the city there were many there who felt his presence.

Martin Image, 2nd plain, jurassic dinosaur producing morrison formation with Book Cliff escarpment on horizon, Utah Highway 24, 1976

An undated Denver Post clipping reads as follows:

Seen with their heads together in Denver this week were Frank Leahy, former Notre Dame coach, and Jack Turner, the fabulous Moab, Utah, uranium claims owner.

Turner handed President Eisenhower a spectacular sample of uranium ore out at Cherry Hills Country Club . . .

A 36 year old native of Moab, Turner was among the real pioneers of the fantastic mid-20th Century uranium rush . . .

We must remember that it was Jack Turner's father who filed the first uranium claim in the state of Utah over eighty years ago, and that Jack Turner is today putting together the proper administrations for what he feels will be one of the largest strikes ever recorded.

The Colorado Springs Gazette Telegraph of September 18, 1955, tells us a bit more about Jack Turner:

Uranium King Acquires Utah Claims From Manitou Group . . .

Jack C Turner, the multi-millionaire uranium king was in Colorado Springs Saturday to close a deal on the purchase of 13 claims in the Red Canyon area of San Juan County, Utah, with some of the sellers being residents of Manitou Springs.

Those who owned the claims from the Pikes Peak Region are the three Higinbotham brothers . . . Two Grand Junction men . . . also were partners. The sale price was not disclosed. Attorney Don Le Mora acted in the transaction.

Turner said that diamond drilling on these claims revealed that in one place uranium ore 12 feet thick assayed anywhere from 1 to 2 percent, which is extremely good ore.

These claims are located close to such famous uranium producing nines as the Gizmo, Maybe and Lasalle.

They are also close in proximity to the Allen Group, now coming into great prominence, where Turner and the Mohler Bros. Enterprises recently drilled into one of the richest discoveries yet found on the Colorado Plateau.

Purchase of the new claims brings the total number controlled by Turner and the companies with which he is associated to more than 1,000 in several western states.

Some of Turner's associates have placed the value of all these claims at about $21 million, but Turner said it is hard to make an estimate of what is in the ground . . . although vast amounts of ore have been blocked out and some stockpiled.

The likeable Turner, who is married and has two children, is a native of Utah and has been in the mining business "most of his life." His home is in Moab.

He got his start with a lease he obtained from the AEC and $15,000 borrowed money. His first good strike was made in TEMPLE MOUNTAIN, Utah, in 1949 . . .

That strike was, of course, upon invitation from Migliaccio.

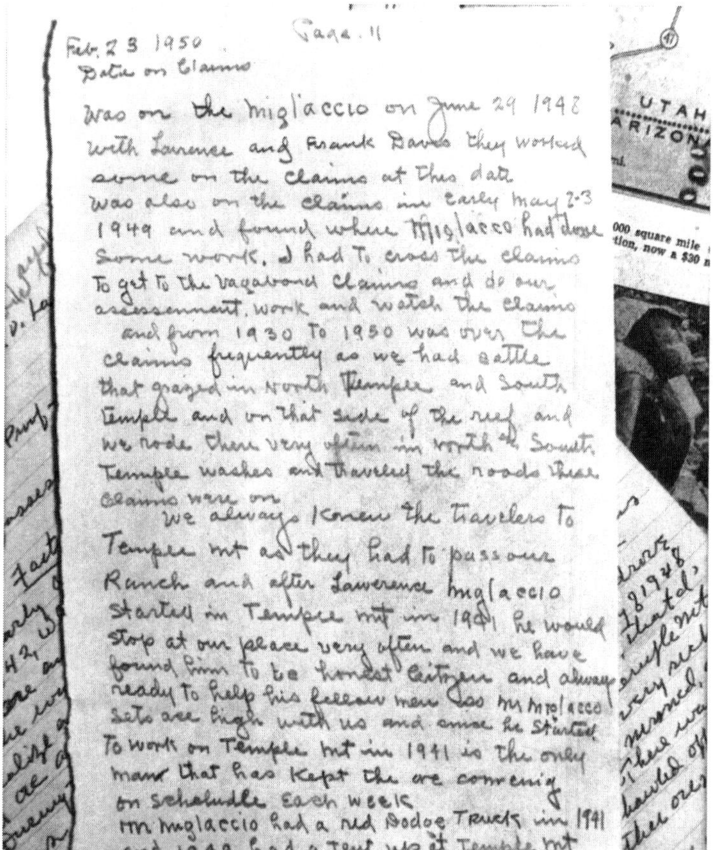

Denny letter graphic.

Martin Image, Leo Cline, Green River, Utah, 1976,
Temple Mountain Miner.

Uranium seekers also touch the persons to descend into Temple Mountains throbbing heart for small halos of uranium ore; entering the earth in 24-inch wide steel shafts in search of radioactivity for the future. Leo Cline of Green River, Utah speaks of entering Temple Mountain for the first time and tells of how he and the BOYS FROM CRIPPLE CREEK . . . hard-rock gold miners from the great mountains of Colorado came to the Sinbad in search of uranium. Leo Cline is one whose days we numbered from too many years too close to the throbbing heart of Temple . . .

". . . he said, "Don't you worry, we're not going to slush this . . . he was talking about a great big slusher bucket that held about ten to twelve tons of ore . . . great big ol' motors on them . . . so we hurried up and drilled; I was working by a guy that everybody said was crazy; I never could see it . . . he was a kinda nice fellow to me anyway . . . well, he was behind, and I was helping him and we were on the last hole . . . he said, you know what? You can never tell what these bosses are going to do . . . they're crazy around here . . . well, that kind of hit me as queer that he would say that . . . he then said, "We are just going to take this rope, we had a starter steel, and a wedge, and put it in this rock so that it will stay . . . I was sitting on a muckpile, you know, glad the day was over with and he went to get five to six sticks of dynamite . . . and just like that . . . the muck dropped out from under me, my hat flew off and my light blew out . . . and down I went . . . I fell to where this curve was and I could feel that rope in my singers and I swung back under a ledge and hung on to my rope and the sparks were just a flyin' . . . I thought, my God . . . if just one rock hits me on the head, I'm dead . . . there I hung and there is no place darker on earth than a mine if you don't have light . . . well, he came back and saw the dust and the muckpile was gone and I was gone . . . he told me later that he said he was goin' down and getting a piece of steel and working that damn foreman over . . . so pretty soon I could hear that slusher running and pulling the muck out . . . trying to find me . . . way down there, see? They found my hat and my carbide light . . . anyway, I got a cramp in my hand . . . I had managed to hang on with one hand and got the rope around my leg a couple of times . . . there I still was . . . in about 35 minutes . . . here they came . . . that old foreman just about got down on his knees to me . . .

Leo Cline glances up for an instant and continues:

"That's what I like about rope, it's saved my life two or three times . . ."

For same time Leo Cline mined Migliaccio Vanadium King claims. Cline has been reputed to be THE MINERS' MINER of the west. There that became involved with him learned from him. Charles Steen and many mining corporations in America have come to Leo Cline for technical help in learning new and more economical ways to follow the ore bodies that ramble out of the gaze of even the most perceptive seeker.

Because of his high regard for the late Lawrence Migliaccio, Mr. Cline has recorded, for the author, the history of Temple Mountain as he knows it, as well as several hours from the most memorable events that occurred in his time with Migliaccio. It must be remembered that Leo Cline was high in the Colorado Rockies, hard-rock mining for gold in the early thirties of this century. As the reader is already fully aware, Leo Cline is a man who has learned the value of ROPE. His vision proves itself of great value when one studies the humanity and not just the administrations and percentage points and percentages of percentages involved in this history.

There are many human beings who have slipped beneath the earth's crust and learned great lessons on existence. The earth is a teacher of man. And Leo Cline has shared with us a few brief instances of the lessons he has learned.

Cline briefly describes his starting in the mining industry:

"... it was hard rock in Colorado in the early thirties ... the hardest rock on earth is in Cripple Creek. The depth of the Depression and no work. Well, a neighbor told me that a relative of hers was looking for partner for mining and I told her that I would try anything once ... well, we were doing pretty well the first time, we had 180 feet to go and a real nice ore body. We went on about 80 feet, then ran into the whole mountain. It had rolled down and rolled down and filled the mine and there was nothing but rocks and dirt a gravel ... we lost about 100 feet of what we were counting on having ... it was a job ... when I first went up there, there were 25 men for each job and like the boss would always say, "If you don't want to put out, there are plenty on top that will!" The superintendent would say that there wasn't a man on earth that was worth five dollars a day ..."

Uranium Seekers

Martin Image, Mrs. Leo, Lillian Cline
first female mayor of Green River, Utah, 1976

Leo Cline was often reminded of the importance of rope, such as when he hung for forty minutes in total darkness as twelve tons of uranium ore rushed past him and twenty-five men waited for him to fall. He is one of the rare breed that is capable of instilling fright in people's hearts with the intensity of an angry glare alone.

Cline is history and legend. Both in the gold capitol of Cripple Creek and in the Temple Mountain mining district, the word of the Clines' was often law. Leo Cline and THE BOYS FROM CRIPPLE CREEK did not always travel alone. Mrs. Leo Cline, one-time mayor of one of the last of the real frontier towns in the west was always with her husband. Her musings upon their fortune, and being a woman mayor in a small town on the fringes of the San Rafael Swell are worth volumes. It must be borne in mind that THE FORTUNE OF TEMPLE MOUNTAIN . . .

is a story of the men and their women who pioneered our west in search of uranium and gold.

Leo Cline recalls instances of his seeing Migliaccio arming himself and going out into the desert blackness to confront "claim jumpers" reputedly sent forth by E. S. Frawley to sabotage the Migliaccio mine-workings. Countless instances of claim jumping and violence have been recorded.

Cline rarely reveals his place in the history until he begins to associate himself with the other persons upon the desert at that time:

". . . guys would come in and spend a half a million dollars on something and have no idea what they were doing . . . when I got started out in Moab I had my own drill rig and even Steen came out . . . it tickled me . . . I was drilling about eighty feet and that was the ore body. He came out and his geologist said it was four hundred and some feet. They set up a Big Rotary Rig and drilled down 450 feet . . . I could never understand engineers you know . . . I showed them where the bedding was. I went down in that shaft . . . 80 feet . . . and you never saw such beautiful ore in your life . . . when you drilled it, it was green . . . oh, the Vanadium in it was awful good. So was the uranium . . . that was $20,000.00 worth of ore . . . the ATOMIC ENERGY COMMISSION man came out with a scintillator and said that there wasn't any ore there . . ."

When questioned as to the affects of living a life close to the creation of all matter; even, perhaps, a little too close, Leo Cline, for an instant, becomes a young man again, hanging by one band in total blindness as 12 tons of ore rush past and he vows that no energy known to man is going to eat <u>him</u> alive from within . . .

Martin Image, Castle Gate, Utah Cemetary, 1976, and the fires still burn

"Did you ever put a spoonful of tar up on a hot roof and then walk on it? You get a gallon . . . that's what I've got all over me . . . dang it anyway . . . I would get it on my rope and my pants and my shoes . . ."

Those who always knew "miners are persons who make light where once there was only darkness."

ACKNOWLEDGEMENTS

Migliaccio Uranium Properties, Brenda Migliaccio Kalatzes, funding 1976-1979, Marsha Craft Cobb, Clyde M. and Ruth Vandeburg, Vandeburg-Linkletter, Associates, Nathan Jones, Laguna Beach, California 1976 historic photo and document graphics, United States Department of the Interior Bureau of Land Management Mining Claims Team, Salt Lake City, Utah and Price, Utah, 1980-2012, Utah State University Eastern Prehistoric Museum, Directors and Staff, Price, Utah, Emery County, Utah Recorder's Office, Castle Dale, Utah, Recorder's, Estelle Guymon, Ina Lee Magnuson, Dixie Swasey and staffs. Stanford J. Layton former Coordinator of Publications and Research, Utah State Historical Society, Clifford Alex Burg, Dallas Shewmaker, Bill Barnett, Ronald Swan, Scot Hahn, Nancy Booker, Jessie Anella and team, Carbon Copy Center, Joel Steinberger, Authorhouse team, Western Mining and Railroad Museum, Helper, Utah, Art and Janet Howard, John Wesley Powell River Museum, Green River, Utah, Museum of the San Rafael, Castle Dale, Utah, Migliaccio Uranium further acknowledges;

Marie O'Neil Migliaccio
C. Nicky Migliaccio-Kalatzes
Michael N. Migliaccio-Kalatzes
Nick Gust Kalatzes
Ellen Rawlings
Lloyd Migliaccio
Tony Protopappas
JoAnne Straight
Diane Nagel
Bob Turri
Racine Turri
Janette Nikas Knowley
Kent Knowley
Polly Bowerman
Vaughn Bowerman
Jacketta World Brewer
Karen Migliacciio Snow
Jay Snow
Tom Migliaccio Sr.
Tommy Migliaccio Jr.
Ronny Migliaccio
Ryan Migliaccio
Todd Migliaccio
Sharon Migliaccio Chipman
Linda Migliaccio Forbes
Carrie Migliaccio Prestwich
Kathy Bullock Turner
Kent Turner
Jim Keogh
Jeff Jacobson
John McFarland
Gary Anderson
Christine Canty Kenny

Duane Matekel
Denise Boyd Kalatzes
Demi Kalatzes
Mia Kalatzes
Sage Copeland
Michelle Madrigal Kalatzes
Gust Kalatzes
Florence Entwhistle Kalatzes
Pamela Davis Austin Reed
Joan Nicholson
Mike Nicholson
Jim Schmitt
Patty Jones Parr
Debbie Migliaccio Howa
Danny Howa
Mark Migliaccio
Brent Migliaccio
Clive Ashworth (Geo Exploration)
John Rud (Geo Exploration)
Dee Openshaw Bradfield
Donald Sheya
Jerri Sheya
Gina Vigil
Holly Johnson
Tom Johnson
Dawn Johnson
John Huefuer
Cleo Huefuer
Linda Gamber Keetch
Brent Keetch
Mara Santi
Michael Howell
Kerwin Jensen

Laurie Milovich
Louie Milovich
Christy Buchanan-Kalatzes
Karle Buchanan-Kalatzes
Dusteen Buchanan
Danny Migliaccio
Lisa Migliaccio
Jack "Cornflakes" Milling, mine caretaker
Migliaccio Sarah M. Mines, Slickrock, Colorado 1950's-1980's

Cretaceous Background. Martin Image,
Historic Helper, Utah, 1976, western railroad and mining town.